An Enthusiasm

for Orchids

AN ENTHUSIASM FOR

Orchids

Sex and Deception in Plant Evolution

JOHN ALCOCK

OXFORD
UNIVERSITY PRESS

2006

OXFORD
UNIVERSITY PRESS

Oxford University Press, Inc., publishes works that further
Oxford University's objective of excellence
in research, scholarship, and education.

Oxford   New York
Auckland   Cape Town   Dar es Salaam   Hong Kong   Karachi
Kuala Lumpur   Madrid   Melbourne   Mexico City   Nairobi
New Delhi   Shanghai   Taipei   Toronto

With offices in
Argentina   Austria   Brazil   Chile   Czech Republic   France   Greece
Guatemala   Hungary   Italy   Japan   Poland   Portugal   Singapore
South Korea   Switzerland   Thailand   Turkey   Ukraine   Vietnam

Published by Oxford University Press, Inc.
198 Madison Avenue, New York, New York 10016

www.oup.com

Oxford is a registered trademark of Oxford University Press

Library of Congress Cataloging-in-Publication Data
Alcock, John, 1942–
An enthusiasm for orchids : sex and deception in plant evolution /
by John Alcock.
p. cm.
Includes bibliographical references and index.
ISBN-13 978-0-19-518274-3
ISBN 0-19-518274-X
1. Orchids—Adaptation. 2. Orchids—Speciation. 3. Orchids—
Conservation. I. Title.
QK495.064A38 2005
584'.4—dc22   2004030008

9 8 7 6 5 4 3 2 1

Printed in China
on acid-free paper

Preface

This book has been written for two groups of readers: those with an interest in evolutionary ideas and those with a special fondness for orchids, especially those that grow in the wild. The evolutionary story as told through the orchid family is colorful, captivating, and full of surprises that will enrich an appreciation for these most appealing of plants.

Evolutionary theory is still very much in the news these days. The big questions in evolution have been debated ever since Darwin by biologists and nonbiologists alike. How can we tell when a feature is an adaptation? How do complex adaptations come into being? Where do species with their special attributes come from? I explore these issues in the pages ahead, explaining how biologists test a feature for its adaptive qualities and explore the origins for such adaptations and how this influences their understanding of how species are formed. These matters are not yet completely resolved, but I do take sides on occasion, knowing that not every botanist and evolutionist will agree with the positions I support.

I turn primarily (but not exclusively) to Australian orchids to illuminate the most interesting evolutionary matters. One could, in theory, use any group of organisms, no matter how small or obscure—British liverworts, Canadian rotifers, the streptobacteria that infect humans, the bracket fungi found on oaks—to provide data relevant for discussions of adaptation and speciation. But there are certain advantages in discussing adaptation through orchids as illustrative matter. First, some orchids are known to have especially intricate relationships with their pollinators, which has affected the evolution of the flowers of these plants. Second, orchids constitute the largest plant family of them all, perhaps as many as thirty thousand species, making up about 10 percent of the world's plant species. This naturally provokes curiosity about the causes for such extreme species proliferation. Thus, orchids provide a rich source of material for an analysis of evolution, adaptation, and speciation.

In addition, many orchids have the advantage of being spectacular plants, so much so that they inspire enthusiasm, even obsession, in some people, making them an especially appealing group to examine. Horticulturists across the globe compete to raise glorious orchids in their homes and greenhouses. A smaller but equally dedicated group like nothing better than to find native orchids growing in the wild, the better to admire them in their natural setting. As it happens, Charles Darwin had connections to both worlds: he praised the exotic species supplied to him by British orchid fanciers and admired Britain's very own orchids as "wonderful creatures."

Some persons are surprised to learn that Darwin had such an interest in orchids given that they have heard so much more about his work with the finches and giant tortoises of the Galapagos Islands, earthworms (whose soil-shifting activities he examined at length), domesticated pigeons (which he kept in order to study the different domesticated varieties), dogs (whose expressions he recorded), and even barnacles, a group that Darwin examined so intensively for so many years that one of his children once asked a friend, "And where does your father study his barnacles?"

But despite Darwin's indisputable interest in animals, he was also a first-rate botanist with a special enthusiasm for orchids. As a young man, he had the good fortune to make the acquaintance of the great botanist John Stevens Henslow at Cambridge University, where Darwin had moved following his abandonment of medicine at the University of Edinburgh. Henslow evidently liked Darwin and took him on numerous botanical field trips, which both men enjoyed. Henslow recognized the scientific potential of his protégé, and it was he who arranged for Darwin to accompany the *Beagle* as the ship's naturalist, where he collected both plants and animals.

When he returned from his travels, Darwin retained a strong interest in plants, and orchids in particular. He and his family often visited a spot they called the Orchis Bank. The area, now called the Downe Bank Reserve, provides habitat for eleven species of orchids to this day. Janet Browne, Darwin's great biographer, suggests that Darwin had this place in mind when he wrote of "an entangled bank" in the last paragraph of *On the Origin of Species*: "It is interesting to contemplate an entangled bank, clothed with many plants of many kinds, with birds singing on the bushes, with various insects flitting about, and with worms crawling through the damp earth, and to reflect that these elaborately constructed forms, so different from each other, and dependent on each other in so complex a manner, have all been produced by laws acting around us" (p. 489).

If Browne is correct, Darwin was thinking of orchids as he completed one of the most important books ever written. In addition, Darwin's liking for orchids led him to write another, much less well known book on these plants, *The Various Contrivances by Which Orchids Are Fertilised by*

Insects. Given Darwin's attachment to orchids, it seems appropriate to focus on these plants when discussing evolutionary matters.

My awareness of the Darwin-orchid connection came after my own fairly recent discovery that orchids are both delightfully complex and varied. For much of my life, I had only a faint awareness of these plants. My earliest recollection of an orchid was the one my mother purchased for me when I was a ninth-grader so that I could give a corsage to Susan Bechtold, my first date ever. However, I do not recall much about the orchid, having spent my time admiring Susan Bechtold rather than studying the corsage that I had nervously given to her at the start of the dance. Later, any potential interest in botany in general, let alone orchids in particular, was stifled by two courses, one taken while an undergraduate at Amherst, and the other as a graduate student at Harvard. My instructor at Amherst, Professor Thomas Yost, taught biochemistry brilliantly, but he abandoned the Socratic method when he was required to teach botany. I don't know about you, but I do not believe that memorizing the difference between parenchyma and sclerenchyma is the most meaningful introduction to botany.

My undergraduate aversion to botany was solidified by Professor Elso Baarghorn, who made graduate-level paleobotany painful by turning on an ancient lantern slide projector at the start of every one of his lectures. In the dark, we were subjected to one black-and-white slide after another of fossil drupes and scapes and bits and pieces of this and that. The experience was excruciating, particularly because the class was taught shortly after lunch, when I was already half asleep.

As a result of my formal education, I at one time concurred with Ambrose Bierce, who defined botany as "the science of vegetables—those that are not good to eat, as well as those that are. It deals largely with their flowers, which are commonly badly designed, inartistic in color, and ill-smelling" (*The Devil's Dictionary*, p. 19). It took a long time and several sabbaticals for me to overcome my antibotanical prejudices. Off I have gone at taxpayer expense to Australia, leaving behind overworked colleagues and ungrateful undergraduates. Instead of committee meeting combat and the pale drama of the lecture hall, I have devoted myself to investigations of Australian native bees, which lead full and active sex lives that are well worth studying. Australia not only has more than enough native bees for American researchers, but it is also a place where English is spoken, welcoming universities are conveniently distributed across the continent, delightful marsupials abound, exquisite beaches tempt the traveler, and automobile traffic is modest to nonexistent (outside of Melbourne, Canberra, Brisbane, Sydney, Adelaide, and Perth). Early on, one or both of our boys came along to admire the kangaroos and thrill to camping in the Australian bush; throughout a whole series of sabbaticals, my wife, Sue (whose maiden name is Coates, not Bechtold), has managed to adapt to life abroad, lending a helping hand with the research and even sometimes receiving the praise she deserved for her assistance.

My 1985 sabbatical was our first to Western Australia, home of Perth and the University of Western Australia, but not much else in the way of human populations and institutions. Although the state of Western Australia covers an area roughly equivalent to the entire western United States, the population of this vast area fails to match the current population of greater Phoenix, where I am just one of several million. The city of Perth, with its pleasant Mediterranean climate and superb beaches, contains the vast majority of all Western Australians, or sandgropers, as they are nicknamed by their countrymen who live in eastern Australia, almost none of whom travel to Perth, as it is considered too distant and isolated. (Incidentally, sandgropers are odd burrowing insects related to the grasshoppers; they move about underground in the sandy soil that covers much of the southwestern corner of Australia.) Outside of sprawling, California-esque greater Perth, one finds great fields of wheat, vast tracts of open space in the semidesert and desert interior, and a few small country towns with a Main Street and a handful of shops, many of which are clearly no longer in business. The largest "cities" outside of Perth in Western Australia, including Geraldton to the north of Perth and Bunbury and Albany to the south, have fewer than thirty-five thousand inhabitants each.

The University of Western Australia occupies land on the edge of the Swan River within view of downtown Perth. Right next to campus the windsurfers race up and down the broad shallow river when the afternoon sea breezes sweep in to make the sport challenging. The Matilda Bay Yacht Club also borders the campus, and almost every day sailboats claim a portion of the river for their races. It is a wonder that anybody in the university gets anything of a nonrecreational nature done. I do not windsurf, however, nor do I own a yacht, and therefore while at the university and in between field projects on bees, I looked around for something of a semirecreational nature that might contribute to my biological education and so be more or less consistent with the terms of my sabbatical, which was granted in the expectation that I would develop new skills while freed from the everyday responsibilities of academe. As I considered my options, someone told me that a team of botanists and photographers were preparing a short expedition to the region to the south of Perth to survey the orchids and certain other native plants of the area. Having nothing else on my agenda, I signed up for the trip with the vague notion that I might see something of interest.

I did not know initially that I would be traveling with two highly accomplished botanists, Andrew Brown and Steve Hopper, as well as two of the top wildlife photographers in Australia, Babs and the late Bert Wells. Sometimes it is enough to be lucky and in the right place at the right time. But thanks to my good luck, I was introduced on this trip to the warty hammer orchid, which completely stole my heart and eventually led me to write this book.

Let me therefore acknowledge Andrew Brown of Western Australia's Department of Conservation and Land Management, to whom I am grateful for introducing me to orchids of southwestern Australia and for providing an especially helpful review of this manuscript under difficult conditions. I also thank Steve Hopper of Kings Park and Botanic Gardens, now of the University of Western Australia, Allan and Lorraine Tinker of Western Flora Caravan Park, and Greg and Mary Bussell, all of whom have helped me find orchids I would otherwise have missed. Let me give special thanks to Babs Wells, who generously provided me with permission to illustrate my book with some photographs taken by her and her husband, Bert Wells. As my enthusiasm for orchids and all that is associated with them grew, I was lucky in having Jim Collins as chair of my department at Arizona State University. He quickly gave me permission to disappear into the Australian bush for months at a time. As noted already, my wife, Sue, has gone along with me, making the long trip west across the Pacific Ocean from Tempe, Arizona, to Perth, Western Australia, many times now. Moreover, she has agreed to live in a cramped campervan for months at a go when we are in Australia. During these periods of enforced intimacy, she has spent many hours wandering through forests, coastal heaths, granite rock reserves, and semidesert scrubland helping me find the orchids that were on my wish list. Our searches have been made all the more productive and pleasant thanks to some Australian friends who have kindly welcomed us into their homes, providing essential relief from campervan living. Win and Ruth Bailey have been especially generous in this regard, as have Tracey and Geoff Allen and Leigh and Carol Simmons. These friends have joined us on assorted field trips, sometimes accepting my decisions on the identity of this or that species, often discovering delightful orchids that I had overlooked. Their good company has magnified my pleasure in developing a better-late-than-never fascination with botany.

Contents

An Enthusiasm

for Orchids

When I saw my very first warty hammer orchid, I was both stunned and delighted. Before me stood a skinny little plant that violated botanical common sense. The orchid had only one leaf, a small blue-green circle of tissue pressed flat against the ground. From the center of this odd leaf rose a thin dark stem about twelve inches high. At the top of the stalk was a weirdly unflower-like flower, the very antithesis of a corsage orchid. The warty hammer orchid flower lacks anything reminiscent of typical petals, and instead features a small purplish cylinder covered with reddish-purple hairs and dark cancerous-looking warts. Were someone to present his prom partner with a warty hammer orchid corsage, I suspect that his date would end even before it began. In fact, I am prepared to concede that the flowers of the warty hammer orchid could be called downright ugly, but I will argue that they are ugly in a most beautiful way.

Once having returned my jaw to its customary position, I looked even more closely at the orchid's flowers. My companion, Andrew Brown, showed me that the warty animalcule could be moved upward in the vertical plane because of a hinge on the horizontal rod to which it was attached. Clearly, the little animal was designed to move or be moved in a particular way, as if it were a machine of some sort. But what on earth was this machine's purpose?

The answer to my question required that I receive instruction on the various parts of a hammer orchid flower and how they relate to those of a normal flower. Andrew told me that all orchids, even the most bizarre ones, such as the warty hammer orchid, have three petals. The hammer

1

Warty Hammer Orchids,

Adaptations,

and Darwin

3

orchid has two thin nondescript ones and a third elaborate and very distinctive lip petal, also called the labellum. In addition, the flower has three sepals, petal-like leaves actually, which are superficially similar to the two true petals that point downward in parallel with the stem. The three petals/three sepals formula characterizes orchids, although this trait also occurs in some other plants, notably the lilies.

The overall design of the orchid flower means that it is bilaterally symmetrical; if one were to cut the flower from top to bottom right down the middle of the labellum, the two halves would be mirror images of one another. A whole other category of plants has radially symmetrical flowers. For these species, *any* division that cuts the flower in half will yield two mirror images, as you can verify by cutting a daisy flower or a cactus bloom in two equal halves. The bilateral symmetry of orchid flowers arises from the specialization of the highly modified lip petal, which has evolved from a standard petal no different from all others into one that can serve as a landing platform for pollinators. The lip petal thus represents an evolutionary innovation that arose after the more generalized, radially symmetrical flower form had originated, presumably because the transformation of a standard, all-purpose petal into a lip petal attracted some especially effective insect pollinators to the plant. The more distinctive the lip petal, the more restricted the list of pollinators attracted to it, and the more likely the remaining specialized pollinators would be to carry pollen from one member of species A to the next, rather than wasting a plant's pollen by donating it to another, different species.

Risa Sargent realized that if bilaterally symmetrical flowers really do tend to have a lock-and-key relationship with a particular specialist polli-

1.1
The warty hammer orchid produces a single flower at the top of a thin stalk; the flower is composed of five short, inconspicuous petals and sepals and one, highly modified petal, the warty labellum.

nator, then any mutational changes in flower structure that spread through a population of a given species would be accompanied by a change in the plant's pollinator. With the switch in pollinators, the affected population would become isolated from the other members of the species to which their ancestors belong, unable to exchange pollen with them reproductively on its own, and therefore, by many definitions, a new species. Thus, the rate of speciation should be higher in bilaterally symmetrical as opposed to radially symmetrical groups, which have far more accessible flowers as a rule, and so generally lack a tight relationship between flower species and one single pollinator. Orchids are bilaterally symmetrical, and many do rely on a single specialist pollinator. Moreover, as mentioned,

the orchid family is loaded with species, so much so that this one group contributes about 10 percent of the entire world's plant species. In contrast, the family most closely related to orchids with radially symmetrical flowers, the Hypoxidaceae, has only a few hundred species. Sargent demonstrated that this same outcome applied to fifteen of nineteen comparisons between families of bilaterally symmetrical plants and their closest familial relative with radially symmetrical flowers. Thus, the kind of flowers exhibited by ancestral plant species may have contributed to the pattern of botanical diversity seen today.

Another important feature of orchid flowers, in addition to their bilateral symmetry, is the column, which is usually composed of one large stamen (the pollen-producing component) and one large style (the pollen-receiving device, which is linked to the carpels, organs where male "sperm" from pollen unite with egg cells to produce the next generation). In the warty hammer orchid, the column is directly opposite the lip petal; the two yellow-green pollen masses form a terminal bulb on this device. Below the pollen, the pollen-accepting stigmatic plate occupies its part of the column. These two reproductive structures, one male and the other female, have fused into a single hermaphroditic club-headed rod in most species of orchids.

The possession of a column separates orchids from almost all other plants, which typically have several to a great many independent stamens resembling miniature lollipops, surrounding one or a few bottle-shaped female structures. In recognition of this difference and some others, such as the minute size of orchid seeds, botanists created the family Orchidaceae. Among the tens of thousands of orchid species are nine different

Warty Hammer Orchids, Adaptations, and Darwin

hammer orchids in the genus *Drakaea*, a group named in the nineteenth century by John Lindley, who wished to honor his botanical illustrator, a Miss Sarah Anne Drake. (I wonder if she was pleased that Lindley had chosen such an odd group of orchids to bear her surname in perpetuity.)

All the multitudes of orchid species have flowers that exhibit variations on a basic theme, albeit very great variations indeed, which is one reason orchids are such a treat. The most distinctive portion of the orchid flower is the labellum, or lip petal, often the key to attracting a particular pollinator. True, in some species, this petal barely differs from the other two, but in many other orchids, such as the hammer orchids, lip petals are strange things indeed. Consider, for example, the bird orchids (in the genus *Pterostylis*) of Western Australia, which include some species with a labellum that is in some ways even more bizarre than the corresponding petal in the hammer orchid. A bird orchid's labellum looks like a collection of short yellow cat hairs glued to a rigid rod a little thicker than a thread, which protrudes from an opening near the base of the hood-like flower. This bizarre device has absolutely no resemblance to what most of us imagine a flower petal to be. The cylindrical flower attracts small flies (gnats and midges), which enter along the labellum, passing through the flower's basal opening. The labellum is counterbalanced and moves upward, closing the opening as the insect crawls down it. Once inside, the flies move upward in the manner of trapped insects everywhere, passing through an upper constriction in the tubular flower toward an opening near the top. As a fly approaches this exit, it may contact the column, brushing any pollen it has acquired elsewhere onto the orchid's stigma before moving past the anther, where it may pick up a new load of pollen to carry to the next bird orchid that the fly chooses to visit.

What an intricate and marvelous piece of floral machinery! And the same can be said for the labellum of the warty hammer orchid, which more closely resembles an insect than a plant part. To understand why, we need to review the mating system of a group of insects, the thynnine wasps, of which there are hundreds of species in Australia. The female thynnine lacks wings, so she crawls up a stem to secure an elevated perch, where she releases a sex pheromone, a chemical attractant, from glands in her head. Male thynnines fly about searching for the odor of a receptive female. When they detect the right pheromone, they fly in a zig-zag fashion upwind until they spot the calling female. They then dive onto the back of the generally much smaller female, sweeping her off her feet and carrying her away for a lengthy copulation. As the pair flies about *in copula*, the male collects nectar from flowering plants and transfers it at intervals to the female, which in some species takes regurgitated fluid directly from her partner's mouthparts. After many such regurgitated meals, the male finally drops his replete mate onto the ground, into which she burrows, carrying the calories she received during her nuptial banquet. Females search underground for beetle grubs, which they sting and paralyze before laying an egg on the victim; the egg hatches into a wasp grub that feeds on the

1.3
In the bird orchids of Western Australia, the labellum has been reduced to a wispy collection of hairs. Insect pollinators enter the opening near the labellum and leave by the upper opening, contacting first the stigma, which will accept pollen carried by the insects, and then the yellow pollen masses, which will donate male gametes to the pollinators as they exit.

flesh of the immobilized beetle. The wasp grub metamorphoses eventually into either a winged adult male that devotes his life to hunting for mates or a wingless adult female that spends as much time as possible tracking down beetle grubs.

Hammer orchids take advantage of male thynnine wasps by producing a scent similar to that of the female wasp. As a result, male wasps are lured to the hammer orchid flower. When they arrive in the vicinity of the plant, they see the labellum, which bears some resemblance to a female thynnine wasp. A male that detects the decoy may dart in and try to fly off with his "mate." As the wasp lifts the decoy, it swings up on its hinged attachment, causing the male to travel upward in an arc that turns him upside down and hurls him into the column. After colliding with this part of the flower, the male, not surprisingly, releases the decoy and attempts to escape. As he struggles to right himself, the back of his thorax may pick up a coat of glue from the upper column. If the wasp then pushes against the anther, his sticky thorax may contact the packets of pollen, pulling them from their chamber. Once the male has recovered his composure and flies off, he carries with him a pollen package glued to his thorax. (The unusual method of packaging pollen masses, or *pollinia*, in a soft or solid clump is yet another distinctive feature of many members of the orchid family.) Should the wasp later be fooled into grabbing another alluring female decoy on another warty hammer orchid, he will crash against that orchid's column, transferring pollen to it when the lumps of pollen on his back are applied to the part of the column (the stigma) designed for their receipt.

More than a million pollen grains may be contained within a single *pollinium*. Some grains are activated after being transferred onto a stigma, growing into tubes that move down through the female part of the column. The pollen tubes carry the plant equivalent of sperm cells into the ovary of the orchid flower, situated below the three petals and

1.4

Thynnine wasps. (top) A female thynnine wasp releasing sex pheromones to attract a mate. (bottom) A winged male thynnine mating with a wingless female; in this species, the male carries the female with him to feed her at intervals with nectar collected at flowering plants.

1.5

Hammer orchid pollination. (top) A male thynnine wasp grasping the female decoy of a hammer orchid (*Drakaea glyptodon*). (bottom) The wasp throws himself into the orchid column as he attempts to fly off with the "female," which is attached to the plant with a hinged stalk. Photographs by Bert and Babs Wells.

three sepals. There the sperm cells unite with egg cells contained within ovarian tissues. Each fertilized egg cell has the potential to become a mature seed. In due course, thousands, even millions, of dust-like seeds are produced in a single ovary to be released after they have matured. A very few seeds may be fortunate enough to land in soil containing threads of the appropriate mycorrhizal fungus, an absolute necessity for seed germination in most terrestrial orchids (see Chapter 7). Should seed and fungus meet up, leading to germination, the resulting plantlet may eventually grow into a new warty hammer orchid capable of producing a wonderful flower in season.

In my sketch of hammer orchid reproduction, I treated the warty hammer orchid's flower as if it were a piece of machinery, each of whose component parts—the labellum decoy with its scent glands, the hinged rod that enables the decoy to move in a particular plane, the column with its gluey pollen masses—seems superbly adapted to achieve a useful goal, namely, the pollination of the plant. This kind of approach has been applied to all living things by observers inclined to view the assorted attributes of all organisms, great and small, as adaptations shaped by natural selection to promote the reproductive success of individuals. The philosophy even has a label, the *adaptationist approach*. Not all biologists, however, adhere to this philosophy. In fact, some have been downright negative about adaptationism, most notably the prominent evolutionist Stephen Jay Gould, now deceased. Gould loved controversy and the hurly-burly of academic debate. You may have read that the reason academicians argue so often and so vehemently is that so little is at stake. Gould would not have agreed. He focused on the big issues in evolutionary biology. Not a shy or retiring person, he did not mince words when lecturing his fellow evolutionists on what he perceived as their shortcomings in understanding central evolutionary concepts.

One such failure, according to Gould, was the widespread acceptance of the adaptationist approach, which he faulted on many grounds but especially because adaptationists insist on viewing living things as if they were an amalgam of adaptations, each one of which might contribute something to an organism's chances of reproducing. Gould believed that, in reality, many traits were less than optimal thanks to various influences, including the effect of evolutionary forces other than natural selection. He claimed that therefore adaptationists were often mistaken in their efforts to tease out the selective value of this or that trait.

Gould made this point and others in one of the most famous papers in evolutionary biology, an article he wrote with his colleague Richard Lewontin, elaborately entitled "The Spandrels of San Marco and the Panglossian Paradigm: A Critique of the Adaptationist Programme." This paper, and others written by Gould, in turn stimulated countercritiques from a number of defenders of adaptationism, among them Richard Dawkins and Daniel Dennett. Even so, the spandrels article has been and continues to be heavily cited by biologists in their research reports, an indication of the respect that Gould and Lewontin's ideas have been accorded. Since the paper was published in 1979, it has received more than fourteen hundred citations, making it a "citation classic." Of these, about one hundred were added as recently as 2003, showing that interest in Gould and Lewontin's arguments continues to climb, even after all these years—not the typical pattern for twenty-five-year-old scientific articles.

In fact, the spandrels paper has become so well known that an entire book was written about it, with contributors from many academic fields deconstructing the meaning of the authors' rhetoric. One of Gould and Lewontin's rhetorical devices was to accuse the opposition of a Panglossian fondness for speculative "just-so" stories about the adaptive purpose of this or that trait, stories supposedly with no more validity than Rudyard Kipling's fanciful account of how the leopard got its spots. Having painted a goodly number of their colleagues in a less than favorable light, Gould and Lewontin really turned the knife in the wound by claiming that Charles Darwin would have agreed that adaptationists were far too committed to natural selection theory than is desirable. In a section of their article subtitled "The Master's Voice Re-examined," they quoted Darwin in support of their claim that this much admired gentleman was not an adaptationist but an evolutionary "pluralist" instead. According to Gould and Lewontin, Darwin may have believed that natural selection was an important evolutionary force, but he also recognized that selection had to share the stage with a variety of other factors that can influence the history of a species. Gould and Lewontin quote Darwin to the effect that he was "convinced that natural selection has been the main, but not the exclusive means of modification" (p. 589) of species. They also highlight this claim: "As far as concerns myself, I believe that no one has brought forward so many observations on the effects of the use and disuse of parts, as I have done in 'Variation of Animals and Plants under Domestication'" (p. 590).

Admittedly, the kind of evolutionary pluralism evident in these quotes posed a problem for Gould and Lewontin, because Darwin was saying that the evolutionary forces he accepted, in addition to natural selection, included a now discredited process of inheritance based on the "use and disuse of parts." Such a process, if it existed, would enable a parent to endow its offspring with parental attributes that had changed during the parent's lifetime because they were used or disused. So, for example, Darwin believed that because burrowing moles rarely exercised their eyes in their underground tunnels, their eyes became weaker, an environmentally induced modification that adults were able to pass on to their offspring.

Darwin's belief in this type of inheritance has considerable psychological appeal and a long history tracing back to the French evolutionist Jean-Baptist Lamarck, who did his work some decades before Darwin. Unfortunately for Lamarckian theory, however, adults that have acquired poor vision through disuse of their eyes, or greater arm strength through use of their arm muscles, simply cannot transmit these modified abilities to their progeny. The fact that Darwin accepted a Lamarckian position reveals just how little was known about genetics and development during the nineteenth century. Thus, Gould and Lewontin acknowledged that they did not "regard all of Darwin's subsidiary mechanisms as significant or even valid" (p. 590) because they knew that Darwin was flat wrong about evolution via the inheritance of acquired characteristics. Nevertheless, Gould and Lewontin quickly went on to say that even though Darwin may have erred when advancing certain of his nonselectionist causes of evolution, nevertheless he had the right kind of "pluralistic spirit," which modern biologists would do well to emulate as an antidote to their Panglossian eagerness to see the best of all possible naturally selected adaptations in every living thing.

But would Darwin have allied himself with Gould and Lewontin? Was he really prepared to put natural selection on a par with assorted subsidiary mechanisms of evolutionary change? Whenever I reread the spandrels paper, I think of the movie *Annie Hall*. In one scene in the film, Woody Allen listens unhappily to a young professor who misrepresents, with great confidence, the views of Marshall McLuhan. Allen says something like, "Wouldn't it be great to have McLuhan himself tell this guy off?" McLuhan then obliges by walking onto the movie set to confront the startled professor, who listens slack-jawed to the corrections he receives directly from the horse's mouth.

Unfortunately, this sort of thing only happens in the movies, as Woody Allen and the rest of us know. Fortunately, we do not need to interview

Darwin today to secure his take on Gould and Lewontin's arguments because he left behind many comments on the subject of adaptation in his books. In fact, Darwin's orchid book, written in 1862 three years after the publication of *On the Origin of Species*, is one of the most adaptationist books of all time. Even a casual reading reveals Darwin's obvious conviction that almost every element of every single orchid flower has an adaptive function, thanks to the pervasive effects of natural selection on the evolution of these plants. Darwin makes this point by pulling apart one orchid flower after another, examining the various components, to figure out how the rostellum of this orchid, the viscid disc of that orchid, and the antennae of still another species advance the pollination chances of the orchid in question. This is precisely what Gould and Lewontin criticize as the supposedly fatally flawed approach of adaptationists, who proceed by "breaking an organism into unitary 'traits' and proposing an adaptive story for each separately" (p. 581).

Consider Darwin's treatment of one of the many British orchids that he examined ever so closely, a species that was called *Orchis pyramidalis* at the time but that has since been renamed *Anacamptis pyramidalis*. You and I can call it the pyramidal orchid in honor of the pyramid-shaped mass of red, pink, or white flowers that run up along the plant's main stem. Darwin knew that this species attracted butterflies and moths. A visiting butterfly inserts its long, thin proboscis into the flower to reach the nectar stored in a long floral spur, which is a highly modified labellar petal. Darwin discovered that an opening to the spur has two ridges that steer the proboscis into the nectary so that the proboscis will contact the pollen-bearing column in the exact manner required to transfer pollinia to the moth. Adaptation number one.

The base of the pollen-bearing device has small sticky flanges that, when touched, quickly wrap themselves tightly around the object that triggered the response. In nature, the triggering stimulus is provided by an

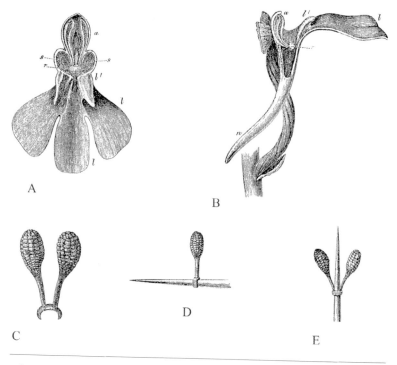

A

B

C

D

E

1.6

Darwin identified adaptations present in the orchid *Anacamptis pyramidalis*. (A, B) A frontal and side view of the flower with the ridged labellum that acts to guide the butterfly's proboscis into the nectary. (C) The two pollen masses as they are mounted on a stalk with a sticky base, which can grasp a butterfly's proboscis. (D) The pollen packets as they are initially carried on a needle (or proboscis) inserted into the nectary. (E) The pollen packets (viewed from above) after they have bent forward in the position that will permit effective pollination of another flower if they are mounted on a butterfly's tongue. From Darwin's *The Various Contrivances by which Orchids are Fertilised by Insects*.

insect's proboscis. During the short time that the insect is removing nectar from the spur, the glue on the flanges sets firmly, so that when the proboscis is withdrawn from the nectary, the pollen-bearing structure comes along as well. Adaptation number two.

At this stage, the moth or butterfly will sport two clubs of pollen mounted at right angles to the proboscis on little stalks held in place by the sticky disc that grips the proboscis. Were the pollen clumps to remain in this position, they would not touch the pollen-receiving surface of another orchid, should the moth or butterfly visit another pyramidal orchid. But after fifteen seconds or so, the stalks automatically bend over so that the pollinia now point forward in a plane parallel to the proboscis. All is in order. When the pollen-bearing moth finds another specimen of the pyramidal orchid and inserts its proboscis into the appropriate channel, the realigned pollen masses will collide with the stigma, thereby effecting pollination. Adaptation number three.

One can see why Darwin exclaimed in a letter to Charles Lyell, the great British geologist, "Talk of adaptation in woodpeckers, some of the orchids beat it" (quoted in Allan, *Darwin and His Flowers*, p. 195). Indeed, Darwin was so pleased to have been able to make sense of several small details of the pyramidal orchid flower, which work together to promote the pollination of the plant, that he even engaged in a little purple prose, speculating that "a poet might imagine that whilst the pollinia were borne through the air from flower to flower, adhering to an insect's body, they voluntarily and eagerly placed themselves in that exact position, in which alone they could hope to gain their wish and perpetuate their race" (*The Various Contrivances by which Orchids are Fertilised by Insects*, p. 79).

In his book, Darwin describes the results of his painstaking dismantling of many more species than the pyramidal orchid. Although he focused heavily on British and European species, he also looked at representatives from all the major groups of orchids, including those found in North and South America, Africa, and Madagascar. The foreign species were often supplied to him by his British friends and acquaintances; he secured some specimens from James Harry Veitch of the famous Royal Exotic Nursery. Having examined all these orchids, Darwin writes near the end of his book, "No one who had not studied Orchids would have suspected that these and very many other small details of structure were of the highest importance to each species; and that consequently, if the species were exposed to new conditions of life, the smallest details of structure might readily be acquired through natural selection. These cases afford a good lesson of caution with respect to the importance of apparently trifling particulars of structure in other organic beings" (*The Various Contrivances by which Orchids are Fertilised by Insects*, p. 287).

Needless to say, Gould and Lewontin did not draw on this quote from the Master in their effort to educate their fellow evolutionists. But none of us need be shocked that Darwin had such confidence that natural selection had influenced even the smallest features of living things. Darwin knew full well that his claim to fame stemmed from his theory of natural selection, not from any vague Lamarckian hand waving nor from his brief comments on other possible influences on evolution. It is natural selection, and natural selection alone, that occupies center stage in *On the Origin of Species by Means of Natural Selection*, and rightly so, because Darwin's theory provided the first convincing mechanism for how adaptive traits might spread through species over time. (Alfred Russel Wallace also came up with the theory of natural selection independently of Darwin and so can be said to share discoverer's rights with Darwin, as confirmed by their simultaneous publication of short scientific papers outlining the theory. But the lion's share of acclaim has always gone to Darwin because he provided much greater evidence on evolutionary processes than did Wallace. To his credit, Wallace never complained that he had been shortchanged, probably because he, like everyone else, admired the depth and rigor of Darwin's science.)

Darwin realized that adaptive change was inevitable if individuals differed in a hereditary trait and consequently also differed in the number of surviving offspring they produced. Such a process will gradually but inexorably modify entire populations as they become dominated numerically by individuals with whatever inherited attribute leads to superior reproductive success. Any trait, no matter how small, that assists individuals in leaving more surviving descendants will become more and more widespread. The extreme adaptedness of living things becomes intelligible once the theory of natural selection is understood.

Moreover, natural selection is the *only* mechanism for evolutionary change that has ever been proposed that is capable of producing the truly complex traits that amaze people and keep many evolutionary biologists at work. Yes, a variety of factors can *interfere* with the evolution of an adaptation, and it's good to know about these things. But when it comes to making sense of developmentally expensive, complex, multifaceted structures, biologists can safely begin their investigations by proposing that these things have an adaptive value that outweighs the cost of their construction. The trick, then, is to figure out what that value might be and then to test one's ideas.

Darwin recognized this point and illustrated the approach when he wrote his orchid treatise. The fact that this was the very first book that he wrote after *On the Origin of Species* tells us something about the special place these plants occupied in his heart. As he says early on in this book, "An examination of [the] many beautiful contrivances [of orchids] will exalt the whole vegetable kingdom in most persons' estimation" (*The Various Contrivances*, p. 2).

When Darwin refers to the "beautiful contrivances" of orchids, he is not speaking of their aesthetic beauty but of the wonder of their Rube

Goldbergian floral mechanisms. Orchid flowers, like the carnivorous leaves of sundews, another subject of Darwin's, are exceptionally complex devices. Indeed, as we have seen already, a good many orchid flowers are so elaborate that one is hard-pressed to recognize them as flowers. They therefore pose a special challenge for the evolutionary theorist. Darwin liked a good challenge because by tackling hard cases, he knew he could demonstrate the universality and power of his theory. To this end, he used selectionist thinking to develop working hypotheses about the adaptive function of each part of the flower.

He increased our understanding of the evolved functions of orchid structures by *testing* his adaptationist hypotheses. When he guessed that such and such a component of the pyramidal orchid had been shaped by natural selection to promote effective pollination, he put his hypothesis at risk. He knew he might be wrong, and he had to find out whether he was. If, for example, the groove on the labellum helped position a proboscis within the flower so that the stalked pollinia could grasp the proboscis, then when Darwin inserted a thin straw into the orchid's nectary via the guides, the viscid disc should attach itself to this object as well. Had the attachment occurred no matter how the straw was placed into the nectary, then Darwin would have been forced to rule against his adaptationist hypothesis for the labellar "guides."

Likewise, when Darwin examined the yellow lady's slipper, he initially thought that pollination occurred when an insect landed on the bowl-shaped lip petal and inserted its proboscis through one of two small openings high on the flower, which happen to be close to a pollen-laden anther. Darwin found that "when a bristle was thus inserted the glutinous pollen adhered to it, and could afterward be left on the stigma" (*The Various Contrivances*, p. 230). However, he also noted that it was hard to deposit pollen on the stigma in this manner, which struck him as odd, and so Darwin was open to an alternative hypothesis suggested to him later by the great North American botanist, Asa Gray. Gray believed that certain insects dropped into the large upper opening on the bowl/slipper, after which they could exit only by crawling upward along the column and then out through one of the little upper openings, a passage that would bring the insect in contact with the pollen, which in yellow lady's slippers is packaged in a sticky fluid.

1.7

The yellow lady's slipper orchid (*Cypripedium calceolus*) with its pouch-like labellum.

Darwin attempted to test Gray's adaptationist hypothesis by dropping some flies into the opening, but "they were too large or too stupid, and did not crawl out properly" (*The Various Contrivances*, p. 230). However, when he introduced a small bee into the labellar chamber, all went exactly as Gray had predicted. The bee could not easily escape from the trap because its smooth, turned-in walls caused the insect to fall back whenever it tried to leave via the large opening above it. Eventually, the bee moved up along the column and out the small opening; when the bee emerged, it had picked up a quantity of pollen. Darwin popped his experimental subject back into the slipper several times, and the bee cooperatively retraced its escape route. When Darwin then dissected the flower, he found pollen smeared all over the stigma, which is located on the escape passage *before* the anther. In nature, this means that a yellow lady's slipper flower will be pollinated only when a bee falls into the labellum carrying pollen acquired from another plant, which will be applied to the stigma *before* the bee picks up new pollen from the recipient flower's anther.

In other words, in this species, as well as most other orchids and other plants, the flower's structure promotes adaptive outbreeding, with plants receiving pollen from other individuals while donating their own pollen to be carried elsewhere, rather than to be used in self-fertilization. Self-fertilizing species of orchids occur in a wide range of genera, including some beard orchids in the genus *Calochilus*, some of which can go either way, being pollinated by insects or fertilizing themselves. Although species like these tell us that *under some conditions* no penalty applies to self-fertilization, nevertheless orchids that use this mode of reproduction make up a distinct minority within the family. In most cases, if a plant were to accept its own pollen, reproducing with itself, the resultant offspring could expect to be genetically handicapped. Inbreeding can result in offspring being endowed with two copies of a damaging form of a gene, one copy having been present in the egg cell and the other in the male gamete. This condition can lead to reduced viability and low-ered reproductive success, as has been demonstrated experimentally with a North American orchid, *Platanthera leucophaea*. When Lisa Wallace took it upon herself to pollinate samples of orchids by hand, she found that seed viability was much reduced for self-pollinated individuals com-pared to those that had received pollen from plants other than them-selves. This is the kind of finding that helps explain why so many orchids have evolved elaborate devices to avoid having their own pollen come in contact with the carpel's pollen-receiving surface. (For example, in one tropical orchid, the freshly removed pollen masses are too large to fit into the stigma, which is a concave pollen-accepting depression on the col-umn; but a half hour or so later, as the pollinia become dehydrated, they shrink so much that they can be easily accommodated within a stigma, which will almost always belong to a plant other than the specimen that provided the pollinia in the first place.)

In any event, Darwin obviously enjoyed his lady's slipper experiments and was delighted to learn later that the bee he had serendipitously used in his study belonged to a genus containing known bee pollinators of another lady's slipper orchid. As a result of the evidence he accumulated, Darwin concluded that his original hypothesis was wrong, whereas Gray's explanation for the unusual features of the yellow lady's slipper was cor-rect—and who would argue with him. Darwin had used the adaptation-ist approach to do good science.

Modern Darwinians, many of whom are adaptationists, have employed fundamentally the same procedures as Darwin in their broad-ranging research. Natural selection theory can be and has been applied to almost everything biological, from why sterile worker ants exist to why people are so committed to religious belief, demonstrating the great scope of the theory. And contrary to Gould and Lewontin, the point of selectionist thinking is not to dream up a "story" about the phenomenon and be done with it, but to think of hypotheses about the potential adaptive value of the trait under consideration in order to test them. Because it is often the case that multiple adaptive explanations for the same attribute are possible, and because some traits indeed are not adaptations, adaptationists must test their ideas. And they do. Without a solid, convincing test, no biological paper on any subject will be accepted by a scientific

1.8

Self-pollination occurs in a few orchids, including some of the beard orchids in the genus *Calochilus*. In self-pollinating species, the pollinia crumble and fall onto the plant's own stigma.

journal these days. No paper, no tenure. It is as simple as that. The drive for tenure and other forms of social success has kept adaptationists hard at work and has resulted in an immense and constantly growing literature on adaptation. This record provides ample evidence that Darwin's adaptationist research program works. Modern adaptationists know, as did Darwin, that even when looking at the fine details, the "trifling particulars" of all living things, not just orchids, odds are pretty good that these traits will prove to be biologically important, to have a function, to bear the imprint of natural selection. This view is as wonderfully stimulating to today's researchers as it was to Darwin, the very first adaptationist. Of course, to put my view to the most severe test, we would have to stage a confrontation between Darwin and Gould, both of whom would require resurrection for the occasion. Where is Woody Allen when you need him?

Most evolutionary biologists who study the purpose or function of one part or another of living things employ the adaptationist approach not because their hero, Darwin, was an adaptationist but because of the long record of success associated with this Darwinian research strategy. As we have seen, Darwin put the method to good use in making sense of the peculiar flowers of many orchids, right down to the "smallest details," the little bits and pieces that make up these extraordinary devices. He did so in part to demonstrate that the many components of a flower could be explained in terms of their contribution to reproduction and as such could be understood as the natural products of natural selection.

2

The Adaptations

of Behaving

Plants

We can illustrate the pervasive utility of adaptationism as a research guide by considering the general phenomenon of plant behavior. Although my academic specialty is animal behavior, I am no chauvinist, realizing that behavior is not the exclusive province of the animal world. Take the warty hammer orchid, for example. The female decoy seems designed to move in a special way after it has been picked up by a male thynnine wasp. True, the orchid's labellum cannot initiate the behavior itself and must rely on male thynnine wasps to propel it during its brief flight. But the orchid labellum has structural features that make sense only in behavioral terms. First, the decoy is shaped, colored, and scented so as to be attractive to male wasps of a certain species. Second, the decoy is placed at the end of a jointed rod with a special hinge. As a result, male wasps can be induced to grasp the labellum and make it move.

The hammer orchid's admittedly limited behavioral repertoire rests on the thin, flexible curl of tissue that occupies the midway point of the labellum rod. This hinge is constructed in such a way that the decoy to which the rod is attached can swing upward. When a male wasp grabs the seductive decoy and attempts to fly away with it, he pivots about the hinge until it is fully uncurled, at which time the insentient decoy and the presumably surprised male wasp have both been turned upside down. In this position, as noted earlier, the wasp's body comes in contact with the column, which is what the orchid needs if pollination to occur.

My point is that the hammer orchid behaves, provided that we define behavior as an *adaptive* movement dependent on complex structures that facilitate the action. When a hammer orchid's labellum moves with the aid of a wasp, it does something useful from the plant's viewpoint. In contrast, when a tree topples in a forest during a violent storm, no one is tempted to label this movement an adaptive behavior because it is evident that the plant gains nothing by falling over.

The hammer orchid is not unique among plants in having body parts that promote adaptive movements. Other examples exist even in Western Australian orchids, including the elbow orchid, whose flower also comes with a labellum that sometimes behaves like the flower of the hammer orchid. I developed a strong desire to see this species behave after reading about it and seeing its picture in *Orchids of South-West Australia* by Noel Hoffman and Andrew Brown (the same Andrew Brown who introduced me to the warty hammer orchid). This quite wonderful book contains a color photograph and a partial page of text for each of the several hundred orchid species found in southwestern Australia. That's right: several hundred. More on the great abundance of Australian orchid species later.

Many good things flow from a decent field guide. Most people want to put a name on things, to have a label for the objects that interest us. A

2.1

The lip petal of the hammer orchid, *Drakaea livida*, has a flexible, hinged rod that enables the orchid decoy to "behave" in a particular way.

field guide's central purpose is to satisfy that desire. On my initial orchid field trip in 1985, I didn't have a book to help me identify the orchids I saw, and even though Brown and his companions gave me names for the species we encountered, these came and sometimes went, leaving incomplete or inaccurate memory traces, sometimes leaving nothing at all. But once I secured the first edition of Hoffman and Brown's book, which I did shortly after coming back from my trip, I spent hours poring over the photographs and memorizing the images and names of species that I had seen or wanted to see.

I was not able to track down the elbow orchid in 1985, but on returning to Western Australia in 1993, I bought the second edition of Hoffman and Brown, which renewed my interest in seeing the elbow orchid and many other Western Australian specialties. True, the elbow orchid is another Australian species with no potential for the corsage or cut flower trade. The entire plant is only about five inches tall at most, with a dirty brown-yellow stem and small rusty flowers that are just as counterintuitive as those of the warty hammer orchid. But for this very reason, I wanted to see the thing in person.

Hoffman and Brown's book gave me hints about how to go about achieving my goal. First, I learned that the orchid does not begin to flower until late October, after most other southwestern Australian orchids have ceased blooming. So I was going to have to be patient (my sabbatical began in January). Second, the orchid book told me that the species grows almost exclusively in shallow pockets of soil found on certain large granite outcrops in this part of the world.

Throughout Australia, the processes of erosion have largely erased whatever topography the continent once had so that the remaining irregularities in the landscape tend to get named. The most famous example of this phenomenon is, of course, Ayer's Rock in central Australia. But the maps of southwestern Australia commemorate a considerable number of smaller, much less visited bumps on Earth's surface: Boyagin Rock, Cardunia Rocks, Queen Victoria Rock, Wave Rock, Kokerbin Rock, Elachbutting Rock, Peak Charles. Most of these outcrops are rounded mounds of pure gray or reddish granite so devoid of soil that they are largely free of plant life. Here and there, however, a thin layer of dirt has accumulated in a gully or little basin carved out of solid stone, providing an ecologically unusual home for certain native plants able to make a lot out of a little. Because agriculture cannot be practiced on these rocks, many have been made into nature reserves in keeping with the world-

wide tradition of ostentatiously conserving those natural areas that have absolutely no economic value. Often completely surrounded by wheat fields or pastures, these granite rocks constitute life rafts for certain specialists, like the elbow orchid, that grow nowhere else on the planet.

I told Bert and Babs Wells, the nature photographers, about my enthusiasm for the elbow orchid, and they pointed me to the unimaginatively named Boulder Rock, a site only about an hour's drive east from Perth in the forested Darling Range. Boulder Rock surges out of the jarrah forest next to a main road so that the rock can be easily reached. Occasional visitors stop for a picnic and have a wander up and over the largely barren rock, sometimes leaving behind a bit of garbage, an apparently universal habit of the human species.

In my eagerness to see the elbow orchid's flowers, I made my initial visit to Boulder Rock in early October, rather than later in the month. As I traversed the naked granite I came to a series of small depressions with a thin layer of soil. I inspected these areas closely on the advice of the orchid book. In no time at all I found the elbow orchid with its unmistakable fleshy mustard-colored stem and little tan basal leaf poking out of the dark soil, which was saturated from a recent shower. The fact that none of the orchids was yet in flower did not greatly disappoint me because I knew that I need merely return in a couple of weeks to see the elbow orchid in its full glory.

When I came back, a few elbow orchids had indeed begun to flower on schedule, coincident with the drying out of the soil, the withering of the basal leaf, and the death of the stem tissue at the base of the plant. Quite remarkably, the elbow orchid reproduces at a time when the plant has begun to expire from the bottom up. Even as the lower part of the stem withers and turns black, flower buds growing higher on the stem begin to open, letting the petals and sepals unfold to reveal a very strange flower indeed.

2.3
Four elbow orchids growing
out of a soil pocket on a granite
rock. The insect-like decoy
flower (close-up) of the orchid
lures male wasps to the plants.
Photograph of the orchid group
by Bert and Babs Wells.

The column of the elbow orchid's flower bends over horizontally so that the clumped pollen are positioned directly above a drooping labellum reminiscent of a hammer orchid's labellum with its hinged rod and scented decoy. In the case of the elbow orchid, however, the rod that connects the decoy to the rest of the flower has a hinge at its base rather than halfway along its length. The elbow orchid's hinge comes into play when a male thynnine wasp (of a different species from the one attracted to the warty hammer orchid) smells the scents wafting from the decoy and accepts the invitation to come for a look. The botanist Warren Stoutamire showed that the odors that do the trick are released from glands in the "abdomen" of the decoy because the experimental removal of the head of the decoy did not eliminate its attractiveness to the male wasps in question. Once a male has approached the labellum, he may, if sufficiently sexually aroused, grab the device and try to carry it away. In so doing, the handsome wasp will travel upward along the constrained route dictated by the hinged rod attached to the decoy before he slams into the column. At this moment, the insect will be more or less horizontal to the ground, and his beating wings will have been trapped by the column's hooked appendages, which fit neatly around his body. It is then that the wasp drops the decoy and struggles to escape from his captor. In his frenzy, the wasp's thorax often presses against the pollinia attached to the column. By the time he finally extracts his wings and flies away, the wasp may be ornamented with a bright yellow mass of pollen, evidence of the adaptive

2.4

Wasp meets elbow orchid flower. (top) A male thynnine is attempting to mate with a female decoy of an elbow orchid. Note that the male already carries a clump of elbow orchid pollen glued to his thorax as a result of a previous interaction. (bottom) After flying up with the female decoy, the male wasp will find that its wings have become trapped in the lateral hooks of the column. In the struggle to break free, the wasp will pick up pollen from or transfer pollen to the orchid. Photographs by Bert and Babs Wells.

The Adaptations of Behaving Plants

value of the various components of the elbow orchid flower. The pollen-laden male may then transfer the plant's gametes to another elbow orchid with a maliciously attractive flower.

So here we have a second case of pollinator-fueled plant behavior in which a male wasp makes a plant part move in such a way as to pollinate the orchid. This kind of plant behavior has been labeled *passive*, in the sense that the plant part does not supply the energy for its behavior (although it does, of course, supply the energy needed for the growth of a labellum capable of attracting a wasp pollinator). Other orchid examples exist, such as an Ecuadorian species in the genus *Oncidium* whose flowers are so delicately attached to the rest of the plant that even a slight breeze sets the petals dancing. These movements may catch the eye of a certain native bee, a species whose males defend a patch of real estate against others of their sex. The territorial male, observing the flowers moving about, may see red and charge. Dashing up to the flower, the aggressive male vigorously rams it with his head, just as if he were head-butting a rival of his own kind to make the emphatic point that he has already claimed this area as his own. Markings on the orchid flower encourage the male to collide with the column. If he does, he will come away with pollinia attached to his head. After the male withdraws from head-to-flower combat, he may transfer the pollinia collected from one plant to another, after a new breeze and another flower shake, in which case the orchids have used the wind to make their flowers move, thereby securing the pollination services of a bee.

One could also argue that passive plant behavior occurs among nonorchids when, for example, a maple seed pod drops from a twig and begins spinning, which prolongs its descent to the earth and helps increase the range of seed dispersal from the parent plant. The structural design of

2.5

The seeds of desert milkweed have delicate feathery tufts; winds pick up the seeds and carry them far from the plant that produced them.

these helicopter seed pods is such that they move in a particular manner, but gravity provides the energy for the movement, not the seed pod itself.

I suppose that the logic of this argument requires that a dandelion or milkweed seed with its parachute is also behaving when a breeze lifts the seed-parachute from the parent plant and carries the device along in a journey to a new location. Clearly, the lightweight parachute is designed to get the seed away from the parent plant, but the energy for dispersal comes from the wind. If we accept these examples as plant behavior, then we can argue against the traditional view, which, as noted by B. S. Hill and G. P. Findlay, has been that plant movements are "rather esoteric events, amusing perhaps, but of little importance in the general biological scheme of things"

("The Power of Movement in Plants," p. 174). To the extent that passive behavior by plants plays an adaptive role in their pollination and seed dispersal, the ability to move takes on great reproductive significance for the species in question.

Moreover, some plants may also "behave" at times when they need to defend themselves, another matter of genuine biological importance. Thus, plants under attack by insect consumers may communicate with others as a means of protecting themselves. When a moth caterpillar, for example, helps itself to one mouthful of leaf after another, it may rupture glandular structures filled with volatile chemicals. These compounds have obvious adaptive value if they deter the herbivore from eating more leaf. In addition, if some of the defensive chemicals become airborne, they can take on a secondary function, which is to attract parasites and predators of leaf-eating caterpillars. If a parasitic wasp were to use a leaf's volatile chemicals to find its caterpillar prey, then in effect the plant has communicated with the wasp, which may be advantageous for both parties. Several researchers have now tested the proposition that leaf volatiles are adaptive signals by showing that when wasp parasites lay their eggs in or on a caterpillar, the parasitized larva sometimes does less harm to the plant than if it were unparasitized.

When a plant benefits by attracting parasites and predators to its herbivores, we can say that the released green-leaf volatiles constitute part of the plant's signal repertoire. Admittedly, communication of this sort qualifies as the most passive kind of behavior imaginable. A considerably more energetic response to herbivory is exhibited by the subtropical tree *Bursera schlechtendalii*. When a leaf of this tree is nipped by an insect, the plant responds by drenching the herbivore with toxic resins that burst from leaf canals, where they have been stored under pressure. Leaves employing this "squirt gun defense" can sometimes actually shoot out a thin but forceful stream of defensive fluid at an enemy. Unfortunately for the plant, this protective reaction can be checkmated by a specialist herbivore, a leaf-eating beetle that carefully severs the leaf's midvein, which lowers the pressure in allied resin canals and thereby untriggers the squirt gun response. The beetle can then feed on the leaves without becoming covered in unpleasant chemicals. However, even though the squirter can be disarmed by the beetle, the ability to discharge resins may still work against some other herbivores. The plant's ability to behave may also force the vein-cutting beetle to spend considerable time preparing a leaf for consumption, thereby reducing the rate of leaf destruction by these insects.

Spraying is an *active* plant behavior as opposed to a passive one because the *Bursera* supplies the energy needed to make the fluid move under the appropriate circumstances. The tree expends costly metabolic energy building up pressure in resin canals so that when a leaf is bitten, the fluid in its canals squirts onto a target consumer. Another kind of active behavioral response by plants has much broader importance, provided we agree that the opening and closing of the tiny pores in leaves (the stomata) is behavioral, even though the movements can be observed only with the aid of a microscope. The leaf pores, or stomata, open and close because of pressure changes within the pair of guard cells that surround each pore. These cells move when they change their turgidity, or internal water pressure, which they can do by regulating the passage of solutes, such as potassium or calcium ions, across their cell walls. When water enters the guard cells following the addition of solutes to the cells, the cell walls expand, causing them to pull away from one another, creating an opening (the stoma) between them. The stomatal opening disappears when two turgid guard cells manipulate their solute concentrations so as to reduce the outward pressure against their cell walls. As cell turgor declines, the two guard cells collapse against one another, eliminating the passage through which carbon dioxide and water vapor can pass.

Leaves control stomatal openings in a very precise and adaptive manner, only permitting carbon dioxide into the leaf when the costs in terms of water loss are manageable. As a result, plants are able to acquire the raw material for sugar manufacture without sacrificing too much of the valuable water they have stored in their cells. The active behavior of the guard cells is therefore central to the energy and water economy of plants, not a merely amusing matter from a plant's perspective but a central adaptation.

I admit that the average person would find watching guard cells in a leaf do their thing an unacceptable substitute for a night of television, even if he or she had access to a microscope suitable for these observations. However, perhaps the jaded "average person" would be somewhat more impressed by other active plant movements that take place on a grander scale than stomatal openings and closings. Hill and Findlay provide many examples of such botanical behaviors, which they further categorize as either *irreversible* or *reversible*.

For an example of an irreversible plant movement, I need only go out into my front yard, where I have planted a small shrub in the genus *Ruellia*. The plant stores its seeds in specially constructed elongate pods. Touch a dried pod and the darn thing literally explodes, sending seeds flying everywhere, the better to get them as far as possible from the parent plant. As the external segments of the intact pod dry, they slowly shrink, placing greater and greater tension on the device. Once the segments suddenly give way, energy is abruptly released, the seeds are shot outward, and the pod's behavior is over, irreversibly finished but with an adaptive outcome: the dispersal of offspring so that they need not compete directly with the parent for light and soil nutrients.

Some orchids also engage in irreversible behavior, with some particularly striking examples to be found among the *Catasetum* orchids of South America. The species in this genus sometimes have unisexual flowers, either male or female, a discovery made by Charles Darwin. Before his work on the genus, male and female flowers of the same species had been classified not just as members of different species but as species that belonged to entirely different genera as well, a fact that reflects the striking differences in appearance between pollen-producing (male) and seed-producing (female) flowers. Orchids in this genus attract large male orchid bees, which harvest strongly scented chemical attractants from the inner lip of the flower's labellum. As the males go about their work day after day, they may accumulate larger and larger amounts of the collected fragrances, which they store in their highly specialized hind legs.

Just exactly why they do so is still not completely clear, although it is known that in some species the odors attract other males, which may gather in groups to display to incoming females. The females may be able to compare potential mates, deriving information about each male's survival and foraging abilities (and thus, genetic quality) by evaluating the quantity and complexity of the scent mixture that he is able to display on demand, a hypothesis that will no doubt be tested some day.

To collect the critical floral fragrances needed for the strange reproductive rituals of orchid bees, the males must crawl deep into the orchid flower. If a scent-gatherer has selected a male flower, eventually his body is likely to touch a thin projection, the antenna, which points downward from the column. When the antenna is pushed, the flower violently fires its pollen-containing apparatus onto the bee's back, where the device sticks firmly, thanks to a gluey base to the packet. For an adaptationist, the question arises, does a plant with the male form of the flower derive a benefit by treating its potential pollen-bearer so roughly? The answer seems to be yes. Clobbered bees learn from a single unpleasant experience to avoid the distinctive male *Catasetum* flower, judging from the inability of observers to find any males carrying two or more pollen packets. By punishing the bee, the orchid manipulates its behavior so that it becomes more likely to carry the plant's gametes to another plant with a female flower rather than to one with a male flower, where it will do its donor no good. And the bees have no aversion to female flowers, which, as I say, do not look like male flowers and do not punish visiting pollinators but instead provide them with rewards, such as fragrant compounds for their hind leg stores. After the male bees have collected these chemicals from a female flower, they attempt to leave. In so doing, they are carefully maneuvered by the flower's structure into depositing the pollen they collected elsewhere onto the stigma, where it can provide the vast numbers of male gametes needed to fertilize the egg cells contained in the flower's ovary.

In reading about *Catasetum* orchids, I came across an article written by Gustavo Romero and Craig Nelson with the claim that the pollinarium of *C. fimbriatum* is shot toward the back of a bee at 323 centimeters per second (almost four feet per second). I wondered how anyone could have come up with such a precise estimate of the velocity of a released pollinarium. Reading on, I learned that the 323 centimeters per second figure came from another paper written in German by a Professor Ebel. I do not know about you, but for me German is not the most welcoming of languages. Long ago, when I was a graduate student trying to stay awake during Professor Barghoorn's lantern slide shows, I was told to learn German if I wished to receive a PhD at Harvard. I made a feeble effort to comply with this requirement but acquired only the most primitive understanding of German, which seemed impenetrable to me at the time. As I peered at Ebel's paper, I could see that nothing had changed over the years. However, even though the text was gibberish to me, the illustrations that accompanied the article made it fairly clear that high-speed photography and frame-by-frame analysis enabled Professor Ebel to produce the 323 centimeters per second result. By artificially triggering the discharge response while filming the orchid and then counting the number of frames between start and finish, Ebel measured how long it took for the pollinarium to go from point A to point B. Then, by dividing the time (a mere fraction of a second) into the distance traveled, which could also be calculated from the film, the author had his estimate in hand. The orchid delivers its pollen via high-speed express.

It cannot be an accident that much of plant behavior has evolved in the context of the interactions between plants and their insect pollinators and prey, which are, of course, capable of rapid movement, thereby

2.6

A foraging plant, one of the Australian dodders whose vines "choose" which stems to parasitize as the vine grows in length.

The Adaptations of Behaving Plants

favoring plants (under some special circumstances) with the capacity to move also. However, other irreversible plant movements take place at a much more leisurely pace than that associated with *Catasetum*. Consider the winding of a vine's tendril around a support of some sort, a form of growth that takes place exceedingly slowly but in a highly directed fashion. Among the many species capable of this sort of thing are certain vining dodders, parasitic plants without roots that take advantage of the metabolic work done by the unlucky host plants they attack. After wrapping around a host's stem like a boa around a rat, the dodder produces small probes that penetrate the host plant tissues, from which they extract energy, nutrients, and water. The botanist Colleen Kelly had the adaptationist expectation that dodders might possess the ability to choose superior hosts to wrap their growing vine tips around, the better to siphon off materials from resource-rich as opposed to resource-poor victims. To test this possibility, she tied dodder stems to hawthorn branches of different

nutritional value, the hawthorns having been grown in the lab with access to differing amounts of a key element, nitrogen. The dodder stems did indeed somehow sense the nitrogen content of the host branch. Branches of "well-fed" plants were usually accepted by the dodder, which looped its stem around the branch one or more times. In contrast, branches with relatively little nitrogen often elicited a rejection response by the parasite, which turned away at right angles from the unacceptable limb, an action that took them about three hours to complete.

In a similar vein, the beach pennywort, a low-growing plant of sand dunes, sends out creeping runners that are somehow capable of veering away from areas covered with competing vegetation. As a result of this ability, the runners more quickly reach open soil patches. There they send down rootlets from which grow new plants; these clonal offspring then repeat the process, colonizing competition-free microhabitats in the dunes in preference to less suitable spots.

To call a pennywort's or dodder's rejection response "behavioral," one has to have tolerance for a time scale that is not usually imagined when one thinks of animal behavior. Charles Darwin not only had the requisite patience to study plant behavior, he began his lifelong investigations into the subject when he was only twenty-four years old as a naturalist on the *Beagle*. While on land in Argentina, Darwin took time out from collecting plants to note the "irritability of the stamens" of a cactus, later named *Opuntia darwinii*. This cactus exhibits a *reversible* behavior that involves the tendency of the clustered stamens to bend toward a bee that is crawling in the cupped flower. The movement of the stamens can be triggered by a gentle touch of a finger or twig, as I myself have seen when playing around with some prickly pear cactus flowers near my home. When these movements have been activated by a bee, they help dust the pollinator's body with pollen, thereby improving the plant's odds of getting its gametes distributed to other flowers.

Later in life, when he was back home in England, Darwin took up the study of plant behavior in a serious way, as documented by Mea Allan in the delightful book *Darwin and His Flowers*. Allan quotes Darwin on this subject to make the point that he viewed plants differently from most of his contemporaries, and indeed from most people today: "It has often been vaguely asserted that plants are distinguished from animals by not having the power of movement. It should rather be said that plants acquire and display this power only when it is of some advantage to them" (from *The Movements and Habits of Climbing Plants*, quoted in *Darwin and His Flowers*, p. 220). Note the adaptationist perspective that Darwin brought to his interest in plant behavior. By looking for the advantages that plants might gain from movement, Darwin catalogued a whole battery of special attributes, such as hooked spines and curling tendrils as well as growth tactics that enable vines and some other plants to climb up to sunlight.

Some modern adaptationists have extended Darwin's approach by making the case that plants are behaving adaptively, not just when a vine's tendril loops around a twig, branch, or trunk, but indeed whenever plants exhibit the flexibility to develop in one way versus another, depending on the circumstances they encounter. For example, many plants adjust both leaf production and root growth in response to the presence of neighboring plants, an adaptability that helps individuals compete for limited sunlight as well as for valuable water and nutrients in the soil. Flexible growth and development also does good things for the wild cucumber. When the plant is young and has limited energy reserves, it produces male flowers (because pollen is relatively cheap to make), but when the plant is older, it then is more likely to generate female flowers, the ones that give rise to the more investment-demanding seeds and fruits. Other plants flower more profusely if they are nibbled by a herbivore during their prereproductive phase (probably because it does not make sense to hold back reserves to live another year if plant-eating enemies are abundant in the neighborhood). Yet another response of an attacked plant is to boost the manufacture of defensive chemicals for its leaves. All these useful "decisions" require that the plants in question have some capacity to sense certain things in their environment and to then change a growth or developmental pattern of some sort. In such cases, it is not out of line to speak of the plant as "behaving."

But the most obvious cases of plant behavior are those in which the plant moves rapidly. One of the more dramatic examples involves the sensitive plant *Mimosa pudica*. Touch a mimosa's long compound leaves and the whole apparatus folds up in a flash. Researchers impressed by the plant's behavior have developed several hypotheses on the benefits mimosas might gain by being so touch sensitive. The rapid folding of the leaf may send an herbivorous insect packing shortly after it has landed on the plant. Alternatively, larger herbivores may be put off by the appearance of the folded leaves, which have a dead or dying look, suggesting that they lack the nutritive value of open leaves. Moreover, when the leaves fold up, they expose the protective spines of the plant. Detailed tests of these adaptationist hypotheses remain to be conducted. In any event, following the departure of whatever it was that activated the rapid folding response, the leaves gradually unfold, so that they can serve the plant again in a photosynthetic capacity.

Other active reversible plant movements take place more slowly than the collapse and reopening of the sensitive plant's leaves. At the very slow end of the behavioral spectrum are the gradual shifts in flower or leaf position that occur in those plants that track the sun over the course of the day. If you have ever been around fields of sunflowers, you will know that the flower heads slowly turn so as to always face the sun, whether it is in the east, overhead, or descending to the west. You may not be familiar with the botanical jargon for this sort of thing, but believe me, there is a term for the behavior: *heliotropism*. (Indeed, there are several forms of heliotropism, but I will spare you the labels.)

Sunflowers are not the only heliotropic plants. Okra qualifies as well, as I noticed in my front yard garden, where the plant's leaves can be observed facing eastward in the morning before gradually turning westward by the afternoon. The adjustments made by okra leaves enable the plant to present the maximum surface area to the sun's rays, presumably boosting their photosynthetic output of sugars. Good experimental studies of a variety of plants other than okra demonstrate that gradual shifts in leaf position during the daytime can increase light capture via solar tracking or decrease damaging heat load and water loss via solar avoidance, depending on the circumstances.

In addition to the solar tracking exhibited by sunflowers, okra, and other plants during the day, some species dramatically alter leaf position at night, a phenomenon with the label *nyctitropism* or *nyctinastism*, depending on your mood and enthusiasm for the seriously byzantine world of botanical jargon. Darwin explored this phenomenon in his book on plant movements, which he illustrated with matched drawings of a *Cassia* during the day, when its leaves are held more or less horizontally to catch sunlight efficiently, and during the night, when they droop downward. Eye-catching plant behavior of this sort led Pliny long ago to comment on the phenomenon. Linnaeus, the author of the modern system of biological classification, published a treatise called *Somnus Plantarum*, not the kind of title to entrance a modern reader but an early venture nonetheless into the world of plant behavior.

Botanists have made considerable progress in identifying the underlying mechanisms that enable plants to fold their leaves into the nighttime sleeping position. A key part of the control system is a biological clock, as shown by the following experiment. Potted nyctinastic plants that have been taken from the outdoors and placed in rooms where the light is on

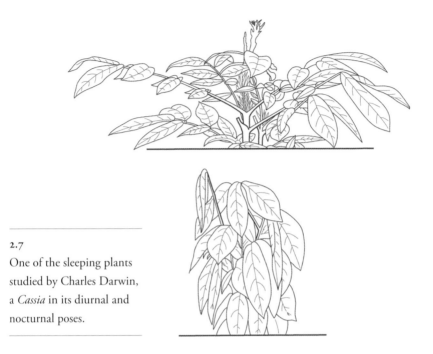

2.7
One of the sleeping plants
studied by Charles Darwin,
a *Cassia* in its diurnal and
nocturnal poses.

continuously (or always off) will for some days thereafter continue to
fold their leaves at about the time when they had been doing so under
natural conditions. The plants then open up their leaves again at about
the time that dawn had been occurring before an experimenter moved
them indoors. Under normal conditions, the hours of sunlight each day
provide a signal that helps set or modify the clock's action, fine-tuning
the control of leaf position. The system is so refined in some species that
not only do the folded leaves open up at the proper hour, but they antici-
pate the direction of the morning sun and begin to shift position before
dawn so as to unfold with the leaf surface turned to the east.

A whole battery of sophisticated components make these daily cycles
of behavior possible. The motor for leaf movement typically resides in

the petiole, or leaf stalk, where it responds to chemical signals regulated by the plant's biological clock. These commands cause one side of a column of cells within the petiole to lose their turgor while the cells on the opposite side of the cylinder retain or gain water pressure. Thus, one side shrinks, and the other side either maintains its dimensions or expands, causing the petiole to bend and the leaf to change its position, sometimes to a great degree.

The elaborate nature of plant biological clocks and hydraulic motor systems underlying their behavior strongly suggests that these devices are adaptations. To my surprise, however, no one seems to know what advantage comes from the ability to put one's leaves to bed at night. Darwin believed that the folded sleeping leaves might benefit from the retention of heat, an explanation he tested by compelling some plant leaves to remain in the daytime unfolded position on cold nights. The experimental leaves did indeed suffer more frost damage than those in the typical nighttime pose.

Freeze protection does not, however, seem to be the reason rhododendrons let their leaves droop and curl during cold days and nights, because leaves in this position are no warmer than erect, flattened leaves. However, leaves in their cold weather position do thaw more slowly than open leaves, which reduces the cellular damage done when icy leaves warm up quickly on sunny days after freezing nights. Moreover, drooping, curled leaves that are exposed simultaneously to strong light and low temperatures on bright, cold days are less likely to incur cellular damage caused by intense radiation.

To return to nocturnal leaf folding, as opposed to wintertime drooping and curling, Darwin's antifreeze hypothesis cannot explain why the behavior occurs in many warm temperate and tropical plant species. These species do not need protection against heat loss and frost damage, and so the whole story behind leaf folding remains elusive. Perhaps in some species, plants close their leaves at night to reduce losing vital ions and other chemicals, which can be washed out of leaves during strong rainstorms. In addition, some have suggested that nocturnal leaf adjustments prevent bright moonlight from interfering with the plant's ability to measure photoperiod (the number of hours of daylight in a twenty-four-hour period, which varies from season to season). For many plants, experimental interruption of a block of nighttime darkness with even a short pulse of light does disrupt the measurement of photoperiod, which in turn can cause the plant to flower at the wrong season. The photoperiod measurement hypothesis, however, has to contend with the fact that many plants with sleeping leaves do not use the duration of photoperiod to regulate important reproductive functions, which leaves us pretty much back where we started.

Although the ecological significance of nocturnal leaf folding is not clear, no such uncertainty attends the reversible slow-motion leaf behavior of sundews (genus *Drosera*) and butterworts (genus *Pinguicula*). The leaves of these carnivorous plants use their ability to move to help them capture and kill their prey more effectively. Both sundews and butterworts have leaves with numerous projections or tentacles topped with gluey glands. Small insects attracted to these glistening fluids become trapped in the glue, which is when the sundews and butterworts swing into action, albeit very, very slowly. In the case of sundews, the relatively long tentacles near the victim begin to bend toward the center of the leaf, adhering to the body of the prey while pushing it inward, eliminating all chance of a Houdini-esque last-minute escape by a struggling insect. Likewise, a butterwort leaf with a frantic fly or bug begins to curl inward, with the leaf edge eventually turning completely over on the prey. This action not only helps restrain captured insects but may also conceal them from potential thieves, such as ants or small birds, that might otherwise be able to pluck the insect from its rightful owner. Indeed, ants often steal exposed prey from carnivorous plants; in one European study, ants promptly made off with nearly 70 percent of the fruit flies that the experimenter had placed on sundew leaves for the sundews' benefit. Carnivorous plants that keep their victims get to digest them and absorb their nutrients, compensating for the lack of key elements in the impoverished soils where they typically grow. Thus, the behaving leaves of insect-eating plants help their owners survive and grow.

Darwin carefully studied sundew behavior before writing yet another of his botanical books, this one entitled *Insectivorous Plants*. Sundews are the most familiar of the six hundred or so plant carnivores, which make up only a tiny fraction of the roughly 250,000 plant species on Earth. Of the roughly one hundred sundew species scattered over the globe, a majority occur in southwestern Australia. There you can find miniature sundews whose ring of tiny leaves is less than an inch across growing near to vine sundews, three to six feet in length, climbing up the trunks of saplings or sprawling over low shrubs. The whole spectrum of sundew types can be encountered during a walk in the jarrah eucalyptus forest near Sullivan Rock, only a short drive from Perth. In late June, tiny microsundews dot the barren clay patches on granite outcrops, and hundreds of pinkish vines festoon the bushes growing on soil patches on Sullivan Rock itself. In the nearby woodland, rosette sundews four or five inches across lie flat on the forest floor, nestled in the litter, each sundew leaf with a handsome green center and red border.

Although Darwin's England was not similarly blessed with multitudes of sundews, two species grew in good numbers on a heath not too far from his home, and so he had a chance to get to know them, which he did not long after publication of *On the Origin of Species*. What he saw convinced him "that *Drosera* was excellently adapted for the special purpose of catching insects, so that the subject seemed well worthy of investigation" (*Insectivorous Plants*, pp. 2–3). As was apparently always the case with Darwin, he threw himself into the study of sundew adaptations with great enthusiasm and interest. His wife, Emma, wrote a friend, "He is treating Drosera (the sun-dew plant) just like a living creature, and I suppose he hopes to end in proving it to be an animal" (quoted in Mea Allan, *Darwin and His Flowers*, p. 98).

2.8

Sundew diversity in southwestern Australia. The sundews differ greatly in size and structure: some species barely an inch tall, others are vines the height of a man.

Emma Darwin was exaggerating, of course, about her husband's desire to prove that sundews were animals disguised as plants, but he was clearly taken by the unusual meat-eating behavior of these plants, as well he might have been, for sundews are awfully good at what they do. Darwin showed that the predatory skills of *Drosera* depended in part on tentacles that were especially likely to move when touched several times in quick succession by a nitrogen-rich substance. The response could begin in as little as ten seconds, hardly a lightning-fast reflex but nevertheless unmistakably behavioral. In contrast, a single touch, even a forceful one, did not do the trick, nor did a raindrop. All of which makes perfectly good sense given that captured insects, a valuable source of nutrients, will struggle for a time to break free, providing the multiple touches that cue the tentacles to respond. In contrast, a raindrop stimulates a tentacle only once. Thus, sundews have a sensory system specifically focused on the detection of prey, which also explains why a bit of raw meat does a better job of eliciting a reaction than an equivalent amount of charcoal, one of

The Adaptations of Behaving Plants

the many, many experiments that Darwin did in the course of cataloguing the special attributes of sundews. At times, his work on this subject seems as obsessive as his investigations of climbing vines; for example, he wanted to establish that a bit of cotton thread weighing all of 1/8,917 of a grain could cause a single tentacle to move if it was placed on the structure with the appropriate multiple touches.

Darwin's interest in carnivorous plants led him to get some specimens of the famous Venus flytrap from a North American contact (the plant grows only in the southeastern United States). As a result, he was able to study this premier behavioral plant at his home in Kent near London. The flytrap responds in less than one hundred milliseconds when a suitable object happens to activate one of six trigger hairs on the two lobes of the trap. These tactile sensory cells send signals that travel widely throughout the lobes, which causes water pressure to fall abruptly in the motor cells that control the opposing components of the trap. The two lobes collapse toward one another so that the spines on their upper edge interlock, loosely at first. If a small and uneconomical prey has triggered the trap to shut, the lucky little insect can walk free through passages between the interlaced spines. In such a case, the trap resets itself, with the plant using its energy

reserves to rebuild water pressure in the relevant cells, pulling the leaf lobes apart. If, however, a large insect has entered the trap, it will be unable to make good its escape as the two lobes then relentlessly press together closer and closer, followed by the release of digestive fluids from glands in the leaves. As the trapped prey liquefies, the leaves take up useful nutrients from the soup.

The capacity for speedy movement has also evolved in some other carnivorous plants much less well known to the general public than the unique and greatly valued Venus flytrap. Among these sprightly behavers are the largely aquatic bladderworts of the genus *Utricularia*, of which there are more than three hundred described species scattered about on every continent except Antarctica. Like the Venus flytrap, they have specialized leaves capable of very rapid movement, which they put to work in capturing prey. In the case of the small and generally inconspicuous bladderworts, the insect-catching bladders consist of small spherical containers with a trapdoor opening. These plants grow in water or water-saturated soils, and so their bladders would be completely filled with liquid were it not for the fact that the plant expends metabolic energy to pump water out of the closed container, lowering the internal pressure relative to the exterior environment. The container walls become deformed as water pressure pushes them inward.

The bladder's trapdoor comes armed with some thin hairs. When a little aquatic invertebrate bumps into one of these sensory hairs, signals are sent to the motor cells controlling the closed door. Shortly thereafter, the container walls lose turgor, the door opens abruptly, and water rushes in, pulling along the prey that had the misfortune to activate the plant's predatory behav-

2.9

Triggerplants behave. The flower on the left has just fired its thin, yellowish column, which flipped forward when the center of the flower was touched. The flower on the right holds its column in the cocked position behind the upper petals, out of view of an incoming pollinator.

ior. The water in the now filled container pushes the door back against the doorsill, eliminating the opening and consigning the prey within to a slow death of the sort Edgar Allan Poe wrote about in *The Cask of Amontillado*. When the prey has at last been digested, the trap can be reset by pumping out the water that accompanied the little animal inward. So here is a fine example of an active, reversible plant behavior.

Rapidity of movement by carnivorous plants makes it possible for them to counter the quick reactions characteristic of their prey, which might otherwise escape. But speedy behavioral responses are not limited to carnivorous plants. Triggerplants, for example, use a nasty little behavioral trick to slap pollen on visiting flies and bees, which come zipping up to the plant's attractive flowers in search of nectar. Should the insect land and begin probing the flower, it may happen to touch the triggerplant's column, which is not identical to that of an orchid but is similar in that

the rod-like structure is composed of both male (pollen-producing) and female (pollen-receiving) components. When the column is cocked and ready to react, it is held behind the flower petals in a bent position out of view of incoming insects. When an insect touches the column's base, the stimulus activates the motor cells in this device so that it flies forward at great speed, traveling between the petals and bringing the pollen-bearing tip down hard on the unsuspecting insect. The fly or bee may be knocked silly in the process, but when it gets its equilibrium back and flies off daubed with sticky pollen, it may be tricked into visiting yet another sadistic triggerplant of the same species. If so, the insect may transfer pollen from the first plant to the other when the column aggressively punches the visitor's pollen-covered thorax.

2.10

Many greenhood orchids, including *Pterostylis sargentii*, have an irritable labellum, which moves when touched. In this individual, the upper flower holds its dark tongue-like labellum in the set position, and the labellum of the lower flower has already shot up into the hood, temporarily trapping the small fly that triggered the flower's behavior.

Thus, it is possible for some plants to benefit from being able to employ an active, rapid response to insect pollinators as well as to insect prey. Which brings us full circle to the orchid world again, where some species get pollinated by being quick on the draw, using their own energy to cock a mechanism that can be triggered by a visiting pollinator. In Western Australia, many orchids of this sort can be found among the greenhoods, shell orchids, and snail orchids that are in the same genus (*Pterostylis*) as the bird orchids we met in Chapter 1. Among these orchids, the dorsal sepal has united with the two adjacent petals to form a cap or hood that shelters the column and labellum. The narrow labellum of many of these orchids can move quickly, as becomes evident when this flower part is touched, even by a mosquito, gnat, midge, or other lightweight fly. When the labellum has been touched, changes occur rapidly in the turgidity of the stem-like base of the labellum, causing it to move suddenly. In some cases, the tongue-like labellum violently jams the fly into the column of the orchid; in other species, the labellum pushes the visiting mosquito, gnat, or midge into the orchid's hood, where it is temporarily trapped. As the small insect searches for an exit, it may come in contact with the column and so may pick up some of the several small granular pollen packets found there. (Because these orchids rely on such tiny pollinators, they do not burden their little fly visitors with large clumps of pollen but instead let the insect carry off a minipacket of gametes of a size appropriate for a small fly.) As usual, cross-pollination requires that the unlucky fly leave orchid specimen number one and make the same mistake of touching the labellum in orchid number two. This must happen with some regularity because certain members of the genus *Pterostylis* are among the commonest of orchids in Western Australia and elsewhere.

The rapid upward movement of the *Pterostylis* labellum is reversible for good reason. For example, should the labellar movement have been triggered "accidentally" by the wind or by contact with something other than a pollinator, the plant can reset the device. The greenhood does so by setting in motion the chemical events that cause the stalk of the labellum to curl downward, pulling the labellum from the chamber where the column is found. Once reset, the device can be triggered again, and if this time the activating stimulus is provided by a midge, mosquito, or gnat, the little fly may find itself captured for a time within the hood, where it may pick up pollen for transport elsewhere—once it has crawled back up to the top of the labellum, past the stigma, and then past the anthers. The labellum then resets itself again, something it can do many times over.

Rather similar behavior is exhibited by another group of Australian orchids, the wonderful flying ducks in the genera *Paracaleana* and *Caleana*. In the flying ducks, the tiny flower has been inverted over evolutionary time so that the labellum, which can be imagined to be shaped like a duck's head, projects upward above the rest of the flower when the labellar stalk has curled under osmotic pressure. This pressure is released when the labellum is touched, causing the labellum to travel rapidly downward so that the duck's bill and head are wedged into a cup-like device called the column wings. In nature, the tactile stimulus that activates the response is sometimes provided by a small thynnine wasp, the male of which may pounce on the labellum as a result of the deceitful sexual odor it provides. If the wasp has firmly grasped the labellum, it will find itself upside down and trapped between the duck's head (the female decoy) and the column. In struggling to get free, the wasp may pick up the column's pollinia for transport to another flying duck.

But sometimes the hair-trigger labellum will fire as the result of a gust of wind or the unintended touch of a human photographer. When this happens, the flying duck orchid then begins to slowly reset its trigger apparatus. As the labellar stalk curls, it pulls the labellum upward and out of the flower's cup into the position needed to attract a male wasp, the key to pollination of the orchid.

As an entomologist, I take pleasure in the fact that speedy insects are responsible for the evolution of plants capable of behaving, especially those species that send pollinators on an unexpected ride or thwack them with pollen or trap them with devices that convert bug into meal. But as

2.11

In flying duck orchids, like this *Paracaleana nigrita*, the labellum is presented to pollinators at the end of a tightly sprung stalk. When the labellum is touched, the stalk relaxes, pulling the "duck's head" downward into a cup, where the pollen-bearing and pollen-receiving column is located. Pollination is effected as the wasp squeezes out of the trap.

a born-again botanist, or at least someone now sympathetic to the study of plants, I am pleased to know that a nervous system is not essential for the detection of certain stimuli and the activation of adaptive responses to an environment. Darwin also noted the similarities and differences between behaving animals and plants in the following passage: "It is impossible not to be struck with the resemblance between the foregoing movements of plants and many of the actions performed unconsciously by the lower animals. . . . Yet plants do not of course possess nerves or a central nervous system" (*The Power of Movement in Plants*, pp. 571, 572). Thus, the parallel abilities of animals and behaving plants are achieved by quite different means, which adds an extra layer of adaptive diversity to the natural world and multiplies the intriguing challenges for those who would make sense of it all.

Thousands of evolutionary biologists ever since Darwin have used Darwinian adaptationism to evaluate competing explanations on how such and such an attribute might contribute to the reproductive success of individuals that possess it. In so doing, they have often been able to demonstrate that the trait in question is in all likelihood an adaptation whose current existence owes much to natural selection in the past. We have seen the utility and power of this approach in our analysis of plant behavior, which is merely one of a great host of biological phenomena that have been profitably examined by adaptationists. This history of success is why so many biologists are convinced that odds are that a given attribute is adaptive, even if the function of the trait is not immediately obvious. Long ago, the evolutionary biologist A. J. Cain expressed this view when he wrote, "It is gradually being realized that if we personally cannot see any adaptive or functional significance of some feature, this is far more likely to be due to our own abysmal ignorance than to the feature being truly non-adaptive, selectively neutral or functionless" ("The Perfection of Animals," p. 37).

3

Adaptations and Maladaptations

Cain backed up his position by showing how an assortment of supposedly nonadaptive traits were later found to have considerable utility after all. But he also acknowledged that some properties of living things cannot possibly be considered adaptations per se. One example that comes to mind is the case of the male thynnine wasp's response to a warty hammer orchid. The male that gets fooled into attempting to mate with the warty labellum of the plant gains absolutely nothing from this moment of sexual overenthusiasm. Instead, the wasp wastes time and energy in an interaction that may well advance the orchid's chances of reproducing but cannot help the male thynnine leave descendants. So here we have a patently maladaptive behavior, a puzzle, a challenge to the adaptationist. I will now argue that, although it might seem counterintuitive, the adaptationist approach can be and has been repeatedly used to resolve these kinds of evolutionary puzzles in a satisfactory manner.

To understand how an adaptationist can explore something that is clearly not adaptive, such as a wasp's eagerness to copulate with a flower petal, we first have to understand what the word *adaptation* means to most evolutionary biologists. Or rather, we first have to understand what it does not mean: an adaptation is not a trait that is perfect, free from all defect, unconstrained by past evolutionary history, unrestricted by the developmental mechanisms that the individual inherited from its parents, and unhampered by the ecological conflicts that the species has with other organisms in its environment. Prior evolutionary events constrain every trait because selection can act only on what is available. Likewise, developmental systems impose limitations on every attribute of every living thing because the genes needed for one characteristic usually code for chemicals that have side effects on the development of other traits. If biologists applied the term adaptation only to those traits that were unconstrainedly perfect in some idealized sense, there would be no adaptations to study.

The definition of adaptation employed by most modern evolutionary biologists is based on the recognition that natural selection cannot construct something out of nothing. Instead, selection is a process that occurs when different hereditary alternatives are present in a population. Whichever alternative happens to be associated with the highest reproductive success spreads the genetic foundation for that attribute, whereas the other types become less common over time as their lower reproductive output gradually takes their genes out of the running. So an adaptation is defined in relative, not absolute, terms. That which is spreading or which is being maintained against other alternatives gets labeled an adaptation, no matter what the constraints imposed on that trait by the species' past history or the interactions between the genes that underlie that trait and other genes that influence the development of other elements of the individual. From the perspective of someone who knows how natural selection works, the expectation is that an adaptive trait will be *better* than other *possible* alternatives, not absolutely without flaw or imperfection.

So when an adaptationist learns that male thynnine wasps will attempt to copulate with hammer orchids, clearly a time- and energy-wasting exercise, he or she might wonder if the male's sexual behavior was better, more effective, than it appears at first glance. Could it be that over time the wasp has evolved some attributes that reduce the obvious disadvantages that come from a tendency to mistake a flower for a female? Those things that improved the wasp's discriminatory capacities could spread through a species even if they didn't enable males to make perfect choices. And indeed, it does not take long to realize that male thynnines are not quite so stupidly copulatory as they appear at first glance, a fact that becomes apparent when one stands around watching flowering hammer orchids. In all my observations of these orchids and their close relatives, I have only *once* seen a male thynnine closely approach a specimen growing naturally in a woodland, and even then the wasp veered away at the last moment rather than grasp the orchid labellum. To observe successful deception of wasp by plant, one usually has to cut the orchid and move the stem and flower to a new location. When this is done, male wasps sometimes, but not always, arrive on the scene quickly, and some, but not all, make contact with the female decoy; a minority of those touching the decoy actually try to fly away with it. In fact, fewer than 10 percent of the wasps approaching experimentally presented specimens of another Australian orchid with a female decoy rather like that of the hammer orchids went so far as to try to mate with the labellum.

On my first orchid-hunting expedition with my Australian colleagues in 1985, we did not use the cut-and-move technique with hammer orchids to see if we could observe the wasp pollinator in action. However, Andrew Brown did try the trick with the zebra orchid, one of the more common species of *Caladenia* in southwestern Australia. The zebra orchid possesses a female decoy, a raised ridge of glandular tissue that occupies the center of the labellum, which looks more or less like a typical flower petal except for the purple-black object lying lumpishly on its surface (see Figure 1.2). When Brown collected some zebra orchid flower stalks, he popped them in a jar of water and kept them fresh in a cooler until he found a spot that he thought might be likely to be home to male wasps of the correct species. But the wasps refused to cooperate, staying away in droves from the orchids even though Brown tried location after location over a period of two days.

3.1
A thynnine wasp perches on the female decoy of a zebra orchid, *Caladenia cairnsiana.*

Finally, when Brown placed the jars in what struck me as a particularly unpromising spot (because of its moderate proximity to a dump with several well-decayed kangaroos that had been deposited there by a local kangaroo shooter), male thynnines answered the call, literally swarming to the jar with the orchids. Many of the males pounced on the flowers as if they had been seriously sexually deprived. Some remained perched on the labellar petal for a half-minute or more, probing the unresponsive tip of the female decoy with their genitalia. Although these males seemed unwilling to take no for an answer, eventually even they gave up. Soon there were no replacements eager to have a go at the decoy. After twenty minutes or so, when all was quiet at the first site, Andrew moved the jar of orchids to a new spot just a few meters away. There the pattern repeated itself, with many males rushing in at first to interact with the decoy, followed by several more minutes of sporadic visits, and then—nothing at all.

When male thynnine wasps are actively searching for mates, they circle through a suitable area over and over again, as I have learned from mark-recapture studies of males belonging to several species. In studies of this sort, I sweep flying males out of the air with my insect net, give them an identifying dot or two of acrylic paint on the thorax, release them, and then record recaptures or resightings of marked males over a period of several hours or days. Recaptures are common, and thus I conclude that males have a small circuit that they travel around, presumably always alert to any fresh source of sex pheromone. When males do smell the scent, they rush to be the first to reach the spot and make off with the female, if indeed it is a female wasp. So why don't the male wasps keep coming time and again to the same orchid flower decoy with its appealing pseudo-sex pheromone?

One answer is that an orchid might modify the bouquet of odors that it produces after it has successfully parted with its pollinia and has received pollinia from another plant. Whether this actually occurs in pollinated hammer orchids is currently unknown, but in a European orchid that tricks a native bee into trying to have sex with a bee-like labellum, the flower's odor does indeed change after pollination has occurred. The change involves a reduction in the release of attractant compounds while the flower simultaneously begins to generate a substance called farnesyl hexanoate. It can be no coincidence that this chemical is also produced by the female bee after she has mated and become sexually unreceptive (having received and stored all the sperm she will need during her short adult life). In interactions among bees, farnesyl hexanoate has the effect of keeping ardent males from pestering already mated females, which has adaptive advantages for both sexes. Females save time and energy in not having to repel unwanted males; the males also come out ahead by not wasting their time pouncing on females that absolutely and positively will not mate with them. By scenting the environment with this very same chemical, a multiflowered orchid also benefits from turning sexually eager bees away from an already "mated" flower. Of course, such an orchid does not invest in farnesyl hexanoate to be helpful to male bees. Instead, by reducing the attractiveness of its already pollinated flowers, the plant may improve the chances that some eager males will turn their attention to the orchid's other, unpollinated, still attractively scented flowers whose pollination will help the plant leave as many descendants as possible.

However, the decline in the number of male insects visiting a deceptive orchid flower might also occur if the pollinators can take matters into their own hands by learning to avoid flowers that trick them into an unrewarding pseudo-copulation. I can imagine two kinds of adaptive learning of this sort. First, a once-deceived male might learn to avoid a particular location, the specific place where he had been fooled into probing a flower petal as opposed to copulating with a female of his species. Alternatively, he might learn to avoid a specific flower, not on the basis of where it was, but by remembering its unique blend of odors, and thereafter avoiding the source of this special bouquet of scents.

The learn-the-location hypothesis is plausible for thynnine wasps given that they do have a familiar home range that they repeatedly patrol and given much other evidence from many insects that individuals can remember the location of distinctive landmarks. Moreover, the hypothesis meshes with observations of males responding avidly to the very same orchids that they had just come to reject—provided that these orchids had been moved a short distance to a new spot. The learn-the-location hypothesis has received additional support from experiments with female thynnine wasps whose mates are attracted to a sexually deceptive Australian orchid in the genus *Chiloglottis*; when a scent-releasing female wasp was placed within a ring of flowering orchids that the local males had learned to ignore, the male wasps also avoided the female, despite the fact that her bouquet of sex pheromones probably was not absolutely identical to that provided by the flowers (a point that needs verification).

But the learn-the-individual-odor-signature hypothesis seems to hold for cases involving some European bees and the sexy orchids that they pollinate. Botanists have analyzed the volatile compounds supplied by female decoys of one such orchid and have found that the labellum gives off more than a dozen different chemical compounds that the male bees can detect. Moreover, the exact amount of compound X generated by a flower varies from plant to plant and even from flower to flower in the same multiflowered orchid. In other words, each flower produces its own unique mix of chemical signals. The male bees respond to this variation by learning the distinctively complex scent of the flowers that fool them. Thereafter, they tend to avoid these familiar odor bouquets, staying away from flowers that they have already visited. In experiments in which a flower was hidden beneath a dead, odorless female bee, males responded avidly to the dummy on the first trial but not on a second presentation in the same location. If, however, on the second presentation of the same dummy at a given spot, a *new* orchid flower was positioned beneath the dead, odorless female bee, the male response to the dummy was just as strong as it had been on the first trial.

An orchid can benefit by providing distinctive labels for each of its flowers because some discriminating male bees with good memories will be able to track a novel scent combination to a fresh flower on a plant with other flowers that male bees have already visited and pollinated. If the bee pounces on the new flower, the plant gets one more chance to reproduce.

The point is that we have learned a great deal about male insect time and energy management by pursuing the adaptationist expectation that these creatures should have evolved ways to reduce the frequency of their mistaken sexual attempts with orchid flowers. The sophisticated ability of certain male bees and wasps to learn things about special odors keeps them from wasting time interacting with flowers they have visited before, with the happy result (from their genes' perspective) that they have more time to find and respond to novel sex pheromone blends, some of which will be produced by receptive females of their species rather than by exploitative orchids. Therefore, male insects that sometimes service deceptive orchids are not totally at the orchid's mercy. Thynnine wasps are far more likely to turn away from the female decoys of orchids at the last instant than they are to fly past a female wasp.

3.2

Orchids in the genus *Chiloglottis* often possess prominent raised calli on the labellum that lure male pollinators to the plant.

Perhaps as a result of waspish discrimination against orchids, pollination rates for many sexually deceptive orchids are low (although infrequent pollination might also stem from a shortage of pollinators in small remnant patches of orchid habitat that have lost most of the necessary insects). In some cases, the pollinator and the deceptive orchid appear to be in an arms race, with the orchids evolving better and better attractant scents and the pollinators evolving better and better discriminating capacity. Because females of each wasp and bee species have their own unique pheromonal blend, deceptive orchids are more or less forced to mimic the lures of one particular species. In this way, the plants have a chance to attract males of one species of insect, which is then enlisted as

the chief or sole pollinator for that orchid species. As a result, a male wasp that has picked up pollen from species A is unlikely to deposit it on a member of species B, where it would not be used. Instead, a wasp carrying pollen in the aftermath of a pseudo-copulation with orchid species A will usually deliver the pollen to another member of species A, should the insect be fooled a second time.

Good evidence for pollinator specificity comes from studies of the species-rich orchid genus *Chiloglottis* in eastern Australia. Orchids in this group are pollinated by male thynnine wasps, that most heavily exploited group of insects, which are attracted to the strange stalked glands called calli on the lip petal. In a study of sixteen species of *Chiloglottis*, Jim Mant and his colleagues found that fifteen were pollinated by male thynnines belonging to the genus *Neozeleboria*. Genetic analyses of the wasps showed that almost all of the fifteen pollinator populations could be genetically differentiated from one another. That is to say, each orchid almost surely had its own species of pollinator, even though most of the wasps were closely related members of the same genus.

The fact that male wasps are generally dealing with just one species of orchid means that selection will tend to favor male pollinators that can analyze pheromonal scents very precisely, the better to distinguish calling females of their species from the orchid that mimics their females' sex pheromone. In the case of at least one pollinator species of *Chiloglottis*, the female wasp's sex pheromone is a single hydrocarbon with the intimidating name 2-ethyl-5-propylcyclohexan-1,2-dione. The orchid also produces this particular substance, which is unique to the plant and its pollinator. In the past, sophisticated olfactory discrimination by the wasp may have selected against orchids that failed to match this special pheromone in every detail, with the result that currently, orchid and wasp match one another extremely precisely with respect to the biosynthesis of a particular molecule.

In other orchid-pollinator relationships, as in the European *Ophrys* orchids that are pollinated by specific *Andrena* bees, the female insect produces a pheromone containing a whole cocktail of scents, favoring individual orchids that most closely match the complete bouquet. The result has been *Ophrys* that produce a mimetic sex pheromone with the whole set of chemicals in place (see Chapter 4).

So perhaps the maladaptive errors made by male thynnine and other dupes of the orchid world (which include males of a few Australian ants and males of a good many European bees) occur because these insects are up against a dynamic, evolving opponent capable in the long term of generating odors absolutely identical to those manufactured by females of the targeted pollinator species. The adaptive features of orchid scent production make it all but impossible for males of certain wasps and bees to avoid making the occasional error. In effect, the orchids are exploiting an otherwise highly adaptive feature of the duped insect's brain. In other words, the mistakes that occur are side effects of a naturally selected behavioral control mechanism that *on balance* maximizes the males' chances of reproduction.

To illustrate this argument we need to consider the nature of interactions between a male thynnine wasp and the receptive females of his own species. The females that are willing to mate are either virgins emerging from the ground after having completed development on beetle larvae paralyzed by their mother or they are adults that have come back up to the surface after an energetically demanding bout of grub hunting underground. In either case, these females typically crawl up a stem and begin to signal for a partner, triggering a competition among the local patrolling males of their species. The first male wasp to reach a female usually gets to carry her off and copulate in return for giving her a considerable amount of nectar. When dropped on the ground, well-fed females can afford to stay underground for some time, out of reach of other males, while they try to find unlucky beetle larvae on which to lay their eggs. Those eggs fertilized by sperm from a recent helpful mate constitute part of his genetic legacy in the next generation.

Given that a male thynnine's reproductive and genetic success is a function of the number of calling females he reaches before any other male, it is not unreasonable to argue that the extreme sexual enthusiasm exhibited by these male insects generates a net reproductive gain for the males in question, despite the time wasted in occasional pseudo-copulatory errors induced by orchids. For the insects, he who hesitates is lost, although perhaps it would be better to say that he who hesitates often loses a chance to pass on his genes. If there were sexually cautious males in the wasp species that pollinates the warty hammer orchid, these males might avoid the alluring scents of the hammer orchid, but they also might well come to a genetic dead end by responding too slowly to signaling female wasps of their kind.

I make this claim with some assurance because of an experiment that a colleague, Darryl Gwynne, and I conducted when Darryl was a postdoctoral fellow at the University of Western Australia. After netting mating pairs of a big thynnine wasp called *Megalothynnus klugii* while the couples visited nectar-producing flowers, we heartlessly pulled male from female wasp in the net. When the female had been disengaged, we slipped her into a black film container for a few minutes, after sending the male on his way. The darkness seemed to calm our captive, which had become agitated (not surprisingly) after being captured and manhandled. When the container was opened and a twig inserted into it, the female usually climbed up the twig a few inches and then stopped, adopting the upright calling position associated with release of a sex pheromone. We inserted the twig into sandy soil and stepped back, stopwatch running. In more than half of sixty such experiments, the perched female succeeded in attracting a male in less than two minutes. We regularly saw two or three males simultaneously tracking the odor source upwind, with the lead male just barely ahead of his rival(s). The *first* male among the many searchers to reach a female was usually able to wrest her from her perch and fly away to copulate with her elsewhere. Therefore, a reproductive premium accrued to the males who were the *most* eager to get to and grab a calling female. The point is that males reluctant to approach the source of their females' sex pheromone are unlikely to leave many descendants that would inherit their sexual reserve.

As noted, the adaptationist perspective is founded on the premise that adaptations, by definition, are those that spread by natural selection because they are better than other, alternative attributes in terms of their contribution to the genetic success of individuals. The traits that persist are not those that are cost-free but those that have a superior ratio of benefits to costs relative to other traits. Thus, a characteristic with some costly effects can nevertheless spread through a population by natural selection provided that this attribute offers a greater net benefit, measured in terms of genes contributed to the next generation, than any alternative characteristic that happens to appear in the species. This understanding has been applied to good effect not just to puzzles in wasp behavior but also to some of the odd features of human behavior as well.

As a behavioral biologist who has done a fair bit of behaving myself, I have had many occasions to wonder why I did such and such instead of this or that, and to regret my failure to do this or that instead of such and such. I will spare you a description of some of my more puzzling decisions, but as I have your attention on matters having to do with orchids, it is not totally out of line to ask why I, and a good many others, are so keen to search for orchids in the wild. (I will leave the fascination horticulturalists have for owning and growing orchids in greenhouses to Eric Hansen, whose book, *Orchid Fever*, deals in part with this kind of orchid mania.) I assume that although a mild obsession with finding orchids in the wild may be somewhat eccentric, it nevertheless is common enough to warrant explanation.

Explaining why we do what we do can be done on several levels. One comprehensive category of explanations deals with the underlying physiological causes of our activities. When we do something, our nervous system is in charge and so is responsible for motivating, activating, and regulating the movements in question, just as the movements of a male thynnine wasp or the leaves of a Venus flytrap require corresponding, but different, physiological foundations. The physiological bases of human behavior have been and continue to be the object of intense investigation, with a constant stream of new findings. Hardly a week passes without the publication of a new magnetic resonance imaging study demonstrating that certain well-defined regions of the brain "light up" with activity when a person craves drugs, views moving lips, sees a fearful face, examines a picture of a landscape, and so on. At its most reductionist, physiological analyses can even focus on how a single neuron, or a given protein, or the gene that controls the production of that protein, contributes to a particular emotion or the ability to do something.

Needless to say, neurobiologists have not found a patch of cerebral cortex, let alone a single nerve cell or gene, whose activity definitely underlies my eagerness to find the warty hammer orchid or any of its many relatives. This is not to say that brain cells and hormones and genes have no causal influences on this aspect of my behavior, merely that we currently do not know just what these influences might be. Although our present ignorance of these matters is unfortunate, on the plus side I am free to speculate about the physiological systems that underlie the orchid hunter syndrome. As a first step toward making sense of my behavior, I have chosen to describe an orchid hunt that I found especially rewarding.

Dragon Rocks Nature Reserve lies toward the northeastern edge of the Western Australian wheat belt, where the rainfall is marginally adequate for the cultivation of wheat in most years, totally inadequate in others, and only truly supportive of the industry on rare occasions. Because of the iffy nature of agriculture in this part of Western Australia and the poor quality of the soils in the reserve, the local wheat farmers never bothered to clear a fair chunk of native vegetation here. The land was therefore available for a set-aside in 1979. Whoever was responsible for the reserve's selection and preservation deserves our gratitude. Just compare Dragon Rocks, a marvelous mosaic of spindly eucalypts and botanically diverse heath, with the surrounding wheatfields, which, depending on the season, are covered with a uniform gray stubble, or green wheatlings, or golden-brown mature wheat, but always a dreary monoculture pure and simple. The contrast between the natural habitat and the agricultural one is painful, especially when one considers that what was once a biodiverse masterpiece has been sacrificed to grow relatively small amounts of low-grade wheat and only in those years when it happens to rain often enough.

But let us be grateful for our blessings, which consist of what is left of the original habitat. And indeed I am grateful. I like the reserve's lack of "improvements," a feature that it shares with many other reserves in Western Australia. No visitors' center, no formal trails, no signs except the minimalist wooden ones that announce the borders of the reserve as you drive abruptly in from wheatfield to bushland on the dusty dirt roads that bisect the reserve. Almost no one ever goes to Dragon Rocks to see the sights, and only a few farmers use the roads through the reserve to get from their isolated homesteads to town now and then.

In the Antipodean spring of 1997 we selected Dragon Rocks as a destination because of its reasonable accessibility and conspicuous presence on our road map of Western Australia. Mapmakers in this part of the world outline the reserves because they have little else to expend ink upon, given the scarcity of towns and roads. I liked the name of the reserve, too, even though the dragons of Dragon Rocks are gray and brown lizards, no more than six inches long, that bask on the gray and brown granite outcrops when it is sunny. Approached by a human, they skitter away at high speed before disappearing beneath flakes of stone that have become partly separated from the mother rock.

After entering the reserve, we drove along, stopping here and there for a roadside wander among the verticordias and starflowers and a host of other flowering plants. Rising out of the low shrubs, patches of impossibly thin-trunked mallees held their terminal tuft of leaves some 5 to 10 meters above the ground. Mallees are a class of eucalypts that send an array of smooth trunks climbing skyward from a central root mass, which can survive even intense fires. The plant regenerates from its root ball after a bushfire, but the plants cannot cope with bulldozers and plows, which tear the roots from the earth. After the land has been cleared, the farmers go along and toss the dense woody balls into the backs of their pickups, or utes, as they are called in Australia. The mallee roots burn for a long time in fireplace or campfire.

As we drove through the reserve, admiring the living mallees, which heightened the aesthetic effect provided by the yellow-, white-, and blue-flowered shrubs underneath the trees, we came across a firebreak cut with surgical precision on an east-west line. The cleared route looked like a good way to get well into the heath without having to fight our way through the prickly shrubs that dominated the landscape. So we moseyed along the firebreak, always scanning, scanning, scanning the vegetated edges of the cleared avenue, occasionally finding something of note: a particularly lovely flowering shrub, some bird orchids, a single curly locks sun orchid, some white spider orchids that had come and gone, their lovely elongate petals and sepals faded and twisted. At each discovery I felt a little jolt of excitement, or what is often spoken of as a surge of adrenalin, reflective no doubt of bursts of neural activity in those parts of the brain that assign pleasure, not pain or indifference, to certain events in our life. Neurophysiologists know something about where and what cell clusters serve to reinforce certain of our activities, making it more likely that we will repeat those pleasurable actions again.

But if my pleasure centers had perked up to some degree whenever I came upon and recognized one orchid or another hiding in the sedges, sheltered under a shrub, poking up through a starflower, or growing brazenly out in the open along the bulldozer-disturbed edge of the firebreak, they really went into high gear late in the trek as we were retracing our steps back to the campervan. The appropriate clusters of neurons went ballistic when I happened to look down at the bark litter surrounding a mallee eucalyptus near the firebreak, a little tree that I had passed on the way out without noting anything special. Now, however, I spotted three small orchids, none taller than 12 inches, each with a single relatively inconspicuous flower. None of the three sepals and three petals was much over an inch long, although they were attractively colored in reds and pale green. The flower featured a prominent labellum that curved forward like a protruding, pointed tongue. I could instantly see that they were members of the genus *Caladenia*, and I thought I knew which species it was, thanks to my hours of field guide study.

I stopped dead in my tracks and dropped to my knees by the orchids, my heart racing just as it had done upon first encounter with the warty hammer orchid. If I had been hooked up to an magnetic resonance imaging device, I have no doubt that the machine would have shown that certain portions, perhaps relatively small portions of my brain, an

3.3
The modest flower of *Caladenia graniticola*, which provided me with great pleasure when I first found it.

organ still precious to me, would have been receiving unusually high rates of blood flow, an indicator of high rates of neural activity at these sites. I guessed that the orchid that was so effective in eliciting a cerebral response was the Pingaring spider orchid, *Caladenia graniticola*, a species found on only a few granite outcrops, according to Andrew Brown and Noel Hoffman.

I marked the spot and dashed back to the campervan, which was not too distant, to check the orchid book and verify my tentative identification, which proved correct. Then it was back to photograph the orchid. As it turned out, I did a miserable job; none of the many photographs proved to be acceptable, which I learned when I saw the developed slides some weeks later. But never mind. At the time I was absolutely on cloud nine.

No doubt one reason I was delighted to see the Pingaring spider orchid has to do with its aesthetic properties, including the colors exhibited by the flowers, especially the lively reds. Bright colors clearly appeal to most people. If evidence is needed, check the flower section of a Burpee's catalogue, which you will find is dominated by large and showy flowers, such as tulips, roses, and other predominantly red and yellow garden flowers. I suggest that the human brain is designed to find red and yellow particularly appealing, emotionally satisfying, desirable, worth a second look, thanks to a visual system that discriminates between colors in the visible spectrum and an allied emotional system that assigns greater emotional value to some colors than others.

All of which raises the question, Why should the human brain have these particular physiological quirks? I mean, why aren't the colors gray and brown as appealing to us as red and yellow? Answers to these questions require something in addition to the claim that the human brain is wired in such and such a manner with neuronal hardware of such and such specifications. The machinery that underlies our perceptions and controls our feelings could be quite different, as shown by the many distinctive specializations exhibited by the nervous system of other animals. There must be reasons why our brains and those of bats, wombats, and gorillas are not the same. And these reasons are evolutionary because the networks of neurons that animals possess have a history heavily influenced by the process of natural selection. Those of us alive today have the brains of the more reproductively successful members of the ancestral populations that preceded us. In the past, humans with brains that gave special emotional weight to reflected wavelengths of light that we currently perceive as the color "red" presumably left more descendants on average than did other individuals who found the color "brown" inherently more conspicuous and more interesting.

But why? Perhaps individuals with a better-than-average capacity to detect reds and yellows found ripe fruits more easily than those who happened to have a somewhat different kind of visual brain. Searchers who discovered red fruit quickly would have acquired the abundant sugars in these ripe fruits, which are often scarce but more valuable than unripe, low-calorie specimens. As a result, these individuals might have consumed more calories on average than their fellows, which in turn could have improved their chances of survival and reproduction. There is no question that many (but not all) of our fellow fruit-eating primates prefer ripe fruits, and so do we, so that perhaps we inherited and then maintained an adaptive neural screening device with a bias for red and yellow colors from an ancestral primate that endowed its descendants with a special visual system.

Some researchers have proposed a different hypothesis for the primate capacity for color vision and the allied ability to identify (and value) red objects. These persons give us reason to believe that special attention to the color red had dietary value for now long-extinct, leaf-eating, rather than fruit-eating, monkeys. In tropical environments, leaves that are red are relatively young, tender, edible, and nutrient-rich, so that the ability to search effectively for reddish leaves would tend to help a foraging animal acquire more useful calories and fewer toxic compounds on the side. Some of the more modern descendants of these pioneering superdetectors of red leaves could well have employed the ability to spot red items when foraging for ripe fruits. If these species were part of the human ancestral lineage, people today have the kind of color vision that carries the evolutionary imprint of both extinct leaf eaters and fruit hunters. All of which may contribute a small part of the total evolutionary explanation for an attraction to red flowers, which is, according to the scenario presented here, essentially an incidental by-product of an element of the nervous system maintained by selection over time for its other, far more adaptive contributions to the foraging success of our ancestors.

3.4

A male buprestid beetle attempts to copulate with the orange reflector shield of our campervan. The dimpled orange shield provides some of the same visual stimuli offered by females of this beetle species.

Evolutionary biologists have long known that neural mechanisms that do something useful can also be employed for nonadaptive or even maladaptive ends. Remember the explanation for why thynnine wasps pollinate hammer orchids? We proposed that this behavior occurs because the male wasps' adaptive mate-detecting systems, which work very well in finding females, can be exploited by orchids outfitted with deceptive female decoys. In some other animals, maladaptive responses occur when a novel feature of the now human-altered environment activates what was once a purely adaptive reaction in ways that do damage to the responding creature. The attraction of moths and nocturnally migrating birds to the bright lights in our cities has led to the death of untold millions of these creatures, which were employing an adaptive orientation response to the light of the moon or stars in an environment very different from the one

Adaptations and Maladaptations

in which the behavior evolved. The ability of a human-altered world to throw an adaptive mechanism out of whack is also evident in the modern behavior of males of certain hefty buprestid beetles in Western Australia. These days the beetles often fly to and attempt to mate with beer bottles and automobile light reflectors that share elements of a color pattern exhibited by females of their species. For 99.9 percent of the evolution of this species, males gained descendants by responding avidly to dimpled orange-brown objects in their environment. But now, with the introduc-

tion of discarded beer bottles and orange reflec-tors on automobiles, the beetles exercise their sexual drive in a totally inappropriate manner, wasting hours and much energy in futile attempts to copulate with certain human artifacts.

In effect, a perfectly sensible visual processing system of the beetle has been hijacked by stim-uli provided by items that the beetles have never had to cope with in the past. Something similar has taken place in recent years in our own species with the development of the car industry and the consequent interest in car identification exhibited by some persons. I remember well my bafflement at the willingness of my two sons to purchase and carefully study automotive magazines when they were teenagers. The information they acquired

from these fine journals and elsewhere enabled them to become awfully good at recognizing the make, model, and age of the vehicles we passed on the street. Judging from a report in *Nature Neuroscience*, they were able to become experts at car identification in part because portions of their cerebral cortex, especially select regions in the fusiform gyrus and occipital lobe, could take on this task. For the vast majority of us, for whom cars are not especially fascinating, these same cortical tissues have a different role to play in our daily lives. Instead of car recognition, they contribute to face recognition, something that almost everyone becomes an expert at during his or her life. Given the highly social nature of the human species and the obvious advantages derived from quickly recognizing friend and foe, family relative and complete stranger, few adaptationists are surprised to learn that we remember faces with great skill. The point is that a neural mechanism that evolved because of its adaptive utility in one biologically significant task can be taken over to perform other, somewhat similar, but far less relevant, activities that were never part of the equation during the evolution of the brain tissue in question.

In a similar vein, our gustatory systems can now be turned on in ways that are almost certainly disadvantageous, thanks to novel alterations in the modern dietary environment of human beings. It is generally believed that people in Western societies consume far too much fatty food today because the human brain tends to interpret the odor and taste of these items in a highly positive manner. A barbecued steak dripping fat onto the coals or a ripe avocado drenched in olive oil or a piece of toast slathered with butter are generally treated as delicacies, even by persons who have no caloric deficiencies whatsoever. The result for many is obesity, with all its attendant social and medical miseries (although a minority view holds that the obesity epidemic stems from overconsumption of sugars and other carbohydrates, which are also available to many humans in quantities greatly in excess of those that could be found and consumed during the evolutionary history of our species).

How to explain the current, probably maladaptive eagerness to indulge in foods high in saturated fats? A physiologist would answer the question by delivering a lecture on the operation of the tongue's taste receptors as well as those brain cells that receive and interpret signals from these receptors, providing us with sensations of pleasure when the food item is that richly marbled steak, but with far less pleasurable perceptions when the person consumes a bowl of oat bran or a fistful of celery sticks. An evolutionist, on the other hand, would provide a different (but completely complementary) hypothesis based on by-product theory. He or she might start with the notion that during the long period when our brains and taste mechanisms were gradually evolving, the risk of eating too much fat was close to nil. Thus, hunter-gatherers whose brains attached special pleasure to the consumption of fatty foods could not possibly have suffered from their preferences but would have been able to discern slight differences in fat content (and thus caloric content) among the generally low-calorie foods available to them. If items with above-average amounts of fat (and calories) were preferred, our ancestors would have been motivated to put more effort into securing foods likely to meet their metabolic needs most efficiently. But the very same mechanisms that achieved this adaptive goal in past environments can in supercaloric modern environments lead to overconsumption of fats, a by-product of our evolutionary history, a legacy of the past effects of natural selection. I am using this same kind of argument when I suggest that my mild obsession for orchids has its evolutionary roots in psychological mechanisms that evolved because they provided useful services of a very different sort in the past. My ancient, long defunct ancestors may have used their color vision for detection of scarce edible leaves or calorie-rich, ripe fruits, whereas I employ this capacity to detect scarce, brightly colored orchids.

Humans have a visual system that not only is tuned to bright colors but that detects and attaches emotional value to *symmetrical* objects, a mechanism that now also plays a nonadaptive role in the appreciation of orchids. The psychological appeal of many orchids and other flowers probably has something to do with the symmetry of these objects. In a typical orchid, the dorsal sepal and labellum form a central component that is framed by the four lateral petals and sepals, which are close to identical in length and appearance. The attractiveness of symmetry of this sort is brought home when one encounters the odd hybrid orchid, some of which have thoroughly asymmetric arrangements of irregular petals and sepals. These hybrids are formed by pollinator errors that occurred when a wasp carted pollen from species A to species B, creating some offspring that carry a mix of genes from the two different species. The development of these individuals is thrown somewhat off course, as the different genes of the two species have not been subject to selection for effective cooperation in the creation of new individuals, unlike the genes that have survived selection within one species. A consequence of imperfect genetic interactions can be imperfect development, as reflected in the lack of symmetry of the flowers of the hybrids, which often lack the aesthetic appeal of the parental species.

But why should human perceptual mechanisms help make visually symmetrical objects more appealing to us than asymmetric ones? Once again we can begin to answer this question by considering the evolutionary reasons for why our physiological systems work the way they do. Being able to detect deviations from symmetry might have had several positive effects during the evolution of human beings. The ability could come into play in the production of tools, if (as I predict) symmetrical tools usually work better than asymmetric ones. But perhaps the main advantage derived from evaluating the degree of symmetry in a visual stimulus that has to do with the selection of a mate. Men and women are highly capable of detecting very small asymmetries in the face and other bodily features. When given an experimental choice between photographs of the same person that have been experimentally manipulated so as to generate different degrees of facial symmetry, people usually identify the most symmetrical image as the one that is most attractive to them, even when the differences between the photographs are remarkably slight. A preference for symmetry in a mating partner has several potential reproductive benefits for the choosy individual. As just noted, genetic deficits of various sorts can result in asymmetrical development, which means that individuals who prefer symmetrical mates might secure better genes for their offspring as a consequence of their preference, without of course having the slightest awareness of the link between their psychological desires and the genetic quality of a possible mate.

In addition, however, asymmetrical development can arise from environmental as well as genetic shortfalls. Diseases, injuries, and inadequate diets, especially if they occur early in life, can alter the course of development in ways that produce superficial asymmetries as well as deeper internal defects. Persons preferring symmetrical mates, therefore, may acquire partners in better physiological condition, who in turn may be better able to bear and nurse children (if female) or provide their partners and offspring with superior protection and resources (if male).

My point, again, is that our physiological capacity for symmetry detection, which may have evolved for one or another useful function, such as the evaluation of potential mates, might be expropriated for orchid admiration simply because orchids coincidentally have the visual properties that activate the symmetry detector. In the same vein, people surely derive pleasure from the odors of certain orchids only because these scents happen to trigger activity in certain parts of our olfactory system, which evolved for reasons other than orchid aroma analysis. Yet the olfactory response to orchids can be strong indeed. No less a personage than Confucius warmly praised the smell of a *Cymbidium* orchid as reminding him of the "joys of friendship," according to Roman Kaiser in his magnificent book, *The Scent of Orchids*. Likewise, persons who wish to sell their house are often told to spray a little canned vanilla scent in the kitchen before the arrival of prospective buyers, the better to seduce their visitors into a deal.

Why should we react so strongly to the aromatic compounds released by some orchids, which manufacture vanilline, limonene, geraniol, methyl benzoate, 1,2,3,5-tetramethoxy benzene, as well as thousands of other chemicals in order to attract pollinators, not the praise of humans? The adaptationist in me proposes that our olfactory peculiarities must have something to do with the fact that we are omnivorous. Humans typically eat a large variety of foods with a correspondingly great variety of volatile chemical components. To analyze potential food items in the hand and in the mouth, we have evolved an olfactory apparatus in the nose, capable of detecting the differences between thousands of aromatic compounds. Moreover, the cerebral component of the smell system attaches emotional valence to the information relayed to the appropriate regions from our olfactory receptors. In other words, we not only can define a scent as almondy, or bell peppery, or musky, or rosy, or fecal, but we find some odors pleasant, others less so, and still others downright offensive. The machinery of smell helps us sniff out and avoid potentially dangerous foods while wolfing down the pleasantly aromatic (i.e., generally nutritious) items we come across. The avoidance of most things rotten, contaminated, or foul makes obvious sense not only in filling out our menus but also simply in terms of steering clear of objects, such as feces, that can contain disease-causing microorganisms.

The analysis of complex scents also plays a surprisingly important role in mate choice. Women find the scent of symmetrical men more attractive than the scent of less symmetrical ones (judging from women's ratings of the attractiveness of odors associated with T-shirts worn by a set of male subjects). Women who are ovulating and thus fertile are particularly likely to give the thumbs-up for the odor of symmetrical guys.

The same T-shirt protocol has also revealed that women can tell the difference between the body odors of men with different MHC genes (a cluster of genes that plays an important role in immune system operation). The women tested preferred the smell of men whose MHC genes were fairly similar to their own MHC genotype, as opposed to those who were identical or not at all similar. If the dislike for odors associated with identical MHC genotypes does contribute to mate choice, it could help steer selective women away from close relatives, thereby avoiding the risks of inbreeding. At the other end of the spectrum, a dislike for odors linked to very different MHC genes could help women avoid males with very different genetic constitutions, thereby reducing the probability of extreme outbreeding, which also carries some risks of producing developmentally disadvantaged offspring. To the extent that women use their scent preferences to select mates with moderately similar MHC genes, they may strike a happy balance between extreme inbreeding and extreme outbreeding to the genetic benefit of their children.

So both the visual and the olfactory system of people have very special attributes that conceivably perform adaptive services for us as reproducing creatures. But these same systems can also be distracted by some incidental stimuli that do not have any direct effect on our reproductive success. And this same point applies to the emotional aspects of human psychology as well as to any visual and olfactory perceptions. Consider my delight at having found a *rare* orchid as opposed to a more common one. Indeed, many common Western Australian orchids far exceed *Caladenia graniticola* in purely aesthetic terms, and yet when I come across one of these lovely, brightly colored, highly symmetrical species my heart fails to pump at the elevated rate it achieved when I found the relatively small, and less obviously beautiful, *Caladenia graniticola* in Dragon Rocks Reserve.

I subsequently found Pingaring spider orchids on another occasion, and the difference between my reactions on the two encounters reveals the importance of novelty as a satisfaction generator. When driving through the wheat belt of Western Australia in 1999, I noticed a massive granite rock outcrop near Pingaring, a very small town that has supplied the common name for the orchid of which I speak. As we drove up to the rock, my wife and I found that it was encircled with a golf course, a primitive golf course admittedly, but a golf course nonetheless. No one was playing on the afternoon we were there, which is not surprising given that only a few hundred people live within fifty miles of the course. But any golf course requires considerable maintenance, and as we wandered toward the granite rock itself, passing first through a fringe of woodland, we came to an open-air depository where the golf course manager had placed the many petrochemicals used to keep the course playable. Whoever was in charge had created a stunning cesspool, which featured decaying containers from which oozed tarry chocolate fluids that positively radiated a black skull and crossbones. I nearly ran back to the van, but decided instead to swing way around the dump and carry on with my search. In due course, I had the pleasure of being surprised to find Pingaring spider orchids in flower growing in the shade of a grove of casuarinas, somber trees with "leaves" that could pass for pine needles. I hurried from the casuarina stand to call Sue to come and admire, which she did, and I cranked up the camera for another round of photography, with somewhat better results this time. But the intensity of my pleasure was a good order of magnitude less the second time around.

Why should that be? The plants were just as lovely as those I found at Dragon Rocks Reserve. But the species was no longer novel to me and so seemed somewhat less rare and unusual than it had on first encounter. Moreover, when I read in my orchid guidebook that the species had been recorded from some granite outcrops in the Pingaring-Newdegate area, I realized that Pingaring Rock had to be one of the known wheat belt sites for the species. In contrast, when I found the orchid at Dragon Rocks, it was growing far from any prominent granite rock in a quite different habitat from the shrub and casuarina groves discussed in the orchid book. In other words, on my first meeting with *Caladenia graniticola* I was able to harbor the conceit that I had found an unusually rare orchid in a place unknown to other, much more experienced orchidologists. (At a later time, Andrew Brown relieved me of this illusion, telling me that the species was in fact known to occur in the Dragon Rocks area.) The second time around I could not entertain the idea that I had made a truly novel discovery that might elevate my status, if only in my own eyes, in the competitive arena of orchid hunting.

One of the universal features of human behavior is the competitive urge. In my circles, I have observed competition at all levels and in almost every possible endeavor, whether it be baseball or darts, amount of income or ability to win at Scrabble, grades received in the classroom or rapidity of advancement up the academic ladder, the number of bird species seen or the rarity of orchids personally discovered. As a young man, Charles Darwin engaged in genteel competition with his peers to see who could collect the most beetle species. Moreover, when older, he wanted to know just how well he had done in collecting the plants of the Galapagos Islands relative to the botanists who had followed him there. His friend Joseph Hooker, who had all the collections in hand, was able to tell Darwin that of the 239 plant species represented in all this material, Darwin had collected 199, of which 180 were new to science. The botanist with the second-best record was responsible for just twenty-one species that no one else had found. Darwin's triumph was overwhelming, and I suspect he was gratified to have come out on top.

Bird watching and orchid hunting share some things with the admittedly rarer beetle- and plant-collecting competitions. All four activities have as a goal the location and "possession" of species previously unknown to the viewer, partly to amass a species list greater than that of others in one's birding or orchid-hunting or beetle-catching or plant-collecting circle. The two top birders I know are constantly reminding one another of the species seen by one but not the other. Just as many birds are difficult to locate (and to identify when seen), so, too, the terrestrial orchids of Western Australia provide major challenges for the searcher. Many species have exceedingly small ranges and extremely tight habitat requirements, which means that they actually occur in only a tiny fraction of the general area where they are reported to grow. In the places where they do exist, these generally small plants often hide in leaf litter or under shrubs or near sedges, where they become all but invisible. Although some of these orchids form fairly large colonies, most occur as scattered individuals. A single specimen here and there can be remarkably easy to overlook.

And in years of drought, which occur regularly in Australia, many orchids refuse to bloom, making their discovery all the more problematic. As I write these lines, I am just back from a day of searching around the town of Mullewa, some 450 kilometers north of Perth. The little patches of native bushland that occur on the roads outside of town provide welcome relief from the wheatfields that dominate the region. They also provide remnant habitat for a variety of orchids—in suitable years. But 2001 has been a year of very little rain, the third-driest on record for Perth. Mullewa is suffering from the drought. The stunted wheat near town is only a foot high, about half what it should be by now. When we stop at a likely orchid site, we find a considerable number of rosy-cheeked donkey orchids, which raises our spirits and gives us unwarranted hope that other orchids might have been able to deal with the drought. But many stops later, our list of species has not grown beyond three, the two other species both represented by fewer than a dozen specimens.

I conclude that finding Western Australian orchids is a challenge and a suitable focus for an oddball competition of sorts. Thus, even though I do not belong to the Western Australian Native Orchid Study and Conservation Group, I have always sought out others to discuss the orchids I have encountered and to encourage some of my colleagues to join me in this activity, in part to show off such skill as I possess in finding these elusive plants. And I was eager to let Andrew Brown, the top dog of Western Australian orchid watchers, know of my "discoveries," in the hopes of establishing my credentials as an orchid finder with him.

The pervasiveness and intensity of competition in human societies has suggested to some that we possess an evolved psychological mechanism that encourages us to engage in culturally acceptable competitions, especially those that we think we might win or at least do well in. Such a mechanism surely has an evolutionary history, and most evolutionary psychologists believe that this history was shaped by reproductive competition. Both men and women have something to gain, reproductively speaking, by being perceived as winners, but more research has been done on male-male competition than on the all-female variety, and so I will focus on the former. In the past, men interested in cultural competitions of all sorts would have engaged in activities that affected an individual's standing with other men. (I trust that it is obvious that I am not claiming that such competitions are good because they are "natural," only that they can potentially be better understood if one has some familiarity with evolutionary theory.)

In human societies, social rank has always had great relevance for a man's attractiveness to women because dominant men typically provide better physical protection and more resources to their partners. Thus, the ability and willingness of men to compete are attributes that presumably evolved because in the past, noncompetitors were often nonreproducers. As a legacy of ruthless sexual selection of this sort, I am endowed with motivational systems that induce me to engage in pursuits as distant from actual sexual reproduction as the search for rare and unusual orchids, the better to impress (I hope) others in my social circle. The odds are minuscule in my case that any gains in social status secured in this manner will translate into additional offspring produced with women awed by my high standing in the orchid search sweepstakes. But even so, it is possible that the genes underlying my brain's operating mechanisms are similar to those of my distant ancestors who chose socially competitive activities more suited to impress rivals and potential sexual partners alike.

One other evolved psychological mechanism also contributes, I believe, to the intensity that some of us bring to the search for scarce orchids, despite the fact that the moments of extreme pleasure are generally few and widely scattered during any given period of hunting. When my wife and I travel through southwestern Australia on one of our orchid jaunts, we stop here and there at spots I judge to be good orchid habitat. Most of our surveys, however, produce little or nothing, no matter how promising the coastal heath or forest reserve appears to be. Even when we have been directed to a locale where our informant assures us that such and such a species occurs, it is almost never easy to find what we are looking for. Recently I was given a little sketch of a map to a site where Cleopatra's needles (*Thelymitra apiculata*) were said to be flowering. This most gorgeous of sun orchids is one that I had never seen, and so I required no encouragement to head to the spot. I followed the directions I had received and arrived at the appointed location, a ridge of orange rocky soil with a ragged coat of small spiny shrubs less than two feet high. I began wandering through the area looking in the fringes of the shrubs for the orchid, which is a strikingly colored species with highly conspicuous flowers. I crisscrossed the heath, but I could not find a single specimen. And yet I kept looking, and finally, forty-five minutes into the search, there the thing was, poking up through a low tangle of twigs that provided the kind of visual clutter into which even a brightly colored flower could disappear. Unfortunately, the flower was mostly closed because the day had been cool and cloudy, conditions that cause most sun orchids to pull their petals and sepals up about the column, a kind of plant behavior that interfered with my desire to photograph a flowering Cleopatra's needle. Even so, I was thrilled just to find a specimen of the plant and so was suitably reinforced for having been a persistent searcher.

One of the most robust conclusions of experimental psychologists and the manufacturers of slot machines is that irregular and unpredictable reinforcement strongly shapes our behavior. We keep doing those things that provide *occasional* rewards, even if these are distributed at considerable intervals. Entire industries—the casinos, the lotteries, the horse and dog racing establishments—are based on the powerful effect that an occasional payoff can have on the person so rewarded. The gambling business has figured out how to exploit this aspect of brain functioning in ways that generate a net gain for the casino owner. The environment of our Pleistocene ancestors, however, was blessedly free from exploitive entrepreneurs with astute accountants and statisticians. In such an environment, the brain mechanisms that now create legions of addicted gamblers and the odd orchid hunter or two presumably motivated our ancestors in ways that, on average, promoted their survival and reproductive success.

One fairly obvious function such a motivational mechanism might have served would have been to encourage the hunter or gatherer of wild foods to keep looking in places where the searcher occasionally made a kill or sometimes found a valuable root or fruit-producing plant. The person who became easily discouraged by a modest run of bad luck might not have persisted long enough to secure a food bonanza that a more dedicated searcher would find. The forager gambling on eventual success could therefore come out ahead of one who did not find the occasional "bingo" so strongly reinforcing that he or she kept looking right through a period of failure.

I suggest that this hunter-gatherer brain mechanism is deeply involved in my own mildly obsessive search for the native orchids of Australia. I am prepared to keep at it for long periods when nothing comes my way because I can remember so well the delight I felt when suddenly, quite unexpectedly, my eye lit upon a "good one" hiding in grasses or peeking out from under the fringe of an overhanging shrub. Just as a hunter-gatherer would gather up the food items found, I also take possession of the rare and beautiful orchids I encounter by photographing them. As a result, I have a huge collection of slides of Australian orchids, the terror of friends and family, most of whom have made it clear that they have had enough of my orchid photographs to last a lifetime. But for me, each viewing provides a little reminder of the excitement, the thrill, the adrenalin surge that came my way when I found my prey and claimed it as my own.

My fundamental argument is that I and other orchid obsessives belong to a species with a history dominated by natural selection, a history that has produced people with highly distinctive brains. The evolved perceptual and motivational systems in our brains had generally useful consequences for our ancestors, whose reproductive success ensured the transmission of their genes, which are around today to influence the development of our cerebral cortex. Our brains, although still capable of adaptive activity, can also generate behaviors of varying degrees of evolutionary novelty and uselessness. As a result, I am wildly enthusiastic about orchid hunting, which is really a by-product of adaptive neuronal systems that did good work for people, including my ancestors, long ago. These same systems now make my own life richer in ways that would seem strange indeed to those of my lineage who, by reproducing, adaptively shaped the evolution of that most complex and marvelous organ, the human brain.

My focus to this point has been on how it is possible to use the adaptationist approach to make sense of both adaptive *and* maladaptive attributes of living things, whether we are talking about orchid flowers or human behavior. Astute readers will have sensed my enthusiasm for this enterprise, for which I am an obvious advocate. But do not think for a moment that evolutionary biology is only about trying to identify the adaptive functions of the features of living things. The discipline is richer than that because it also deals with another extremely intriguing issue in addition to adaptation: How did this or that attribute come into being? What, for example, is the history behind the elaborate flower of the warty hammer orchid (*Drakaea livida*)? What did the first orchid flower look like, and what were the series of changes that took place in its modification over evolutionary time that eventually resulted in hammer orchid flowers in one lineage of this group of plants? Knowledge of the reproductive functions of the warty hammer orchid flower cannot substitute for information about the history of changes in this species' lineage.

4

The History in Evolution

We alluded to the history of a trait when discussing the possible origins of human color vision in a distant fruit- or leaf-eating primate ancestor. Darwin provided a way to reconstruct cases like this with his theory of descent with modification, which complemented his theory of natural selection. Darwin's descent theory is based on the assumption that the complex features of modern species have their beginnings in the simpler traits of long extinct species. Imagine an ancestor of an existing species, one with a much earlier form of a characteristic exhibited by a living organism. When this ancestral species split into two populations, the stage was set for the eventual evolution of two distinct descendant species, each initially endowed with the hereditary information of their immediate predecessor. As a result, these separated populations at first possessed features that were identical or very similar to those of their immediate ancestor. As time passed, however, genetic differences would begin to accumulate in each line as each population evolved under the influence of natural selection acting on the different mutations each happened to receive. By the time these populations became "new" species and split to give rise to their own descendants, they would probably retain some shared features, but in addition would differ thanks to the genetic changes that had built up in each lineage.

As the process of descent with modification repeated itself time and time again, the more recent products would become less and less like the distant ancestral species at the base of a branching tree. By the time we reach the present, with its current set of species, none of these organisms would likely have more than a residual resemblance to any ancestor located far back in time. Understanding the origins of the special attributes of a current species therefore requires that we somehow climb down the tree of life, starting with an outer twig and working our way back along the branches and trunk to the base of the tree, unlayering the modifications that have accumulated in an evolutionary lineage over time.

One way to do reverse tree climbing of this sort is, of course, to find the appropriate sequence of fossils, correctly dated so that one can go back in time, working one's way back from the near present to the distant past, making the connections that enable one to write a history of a lineage. One could conceivably employ this strategy in outlining the evolutionary changes that are responsible for the evolution of something like the warty hammer orchid. Plants, as well as animals, can turn to stone, as witness the remarkable fossilized trees that I and millions of others have admired in Arizona's Petrified Forest National Park, not that far from my home. Even delicate plant parts, including flowers, can fossilize under some special circumstances, and these fossils have helped botanists test ideas on the evolutionary origins and subsequent diversification of the flowering plants. Indeed, not too long ago, a paleobotanical team reported that they had found a possible very early ancestor of today's flowering plants. The nicely preserved specimen, only about a foot tall, was removed from rocks dated to 125 million years before the present, a time that long preceded the explosion of flowering plants, which started in earnest about 65 million years ago. Inspection of the fossil impression revealed that the plant, now named *Archaefructus*, possessed the two essential elements of an angiosperm or flowering plant: pollen-bearing stamens and fruit-containing carpels. You will remember that the carpel is the jug-like female reproductive structure that contains the fruits after they are formed. The plants that existed before the angiosperms, the conifers and other gymnosperms, produce seeds that are not enclosed within a fleshy structure.

As one might expect from a "primitive" original flower, the carpel and stamen of *Archaefructus* were not advertised by petals and sepals, and so this flowering plant would not have been much to look at. You can be sure, however, that Ge Sun, David Pilcher, and the rest of their research team were hugely pleased by the fossil remains of the plant because it enabled them to test and reject one plausible and widely held theory of plant evolution, namely, that the ancestral flowering plant was a tree rather like a magnolia. Their find, *Archaefructus*, was instead a small weedy plant, probably a swamp dweller.

The last word has not been written on the origins of flowering plants, a statement that applies with even greater force to the origins of orchids. To date, almost nothing has been written on the history of orchids as revealed by plant fossils for the simple reason that such fossils are exceedingly rare. Even the very few fossils that have been assigned to the orchid family by one authority have been challenged by others, thanks to the fragmentary and confusing nature of these materials. The scarcity of good fossil orchids is not all that surprising because most ancestral orchids were surely small, soft, and delicate (like most modern orchids), and therefore prone to quick decay and disappearance upon death. Presumably, the immediate ancestors of the warty hammer orchid were similar in size and shared a preference for dry, sandy woodlands and so were unlikely to persist long after death. Fossilization generally requires that organisms avoid decay for some time, which is much more likely to occur in aquatic environments where a dead animal or plant can become quickly covered with preservative silty deposits.

Nevertheless, even when fossil evidence is lacking, one can still evaluate some hypotheses about the history of a group by making use of data provided by *living* species. For example, if we want to get a rough idea of when the orchid family originated, one way to go about it is to look for clusters of related species that occupy what were once parts of Gondwana, the ancient megacontinent whose breakup yielded today's South America, Africa, and Australia. Because South America and Africa pulled away from one another on the order of 100 million years ago, the discovery of closely related orchids on the two continents today would suggest that their common ancestor occupied Gondwana. If so, orchids would have to be at least 100 million years old. Modern orchids in the subfamily Vanilloideae (which includes members of the famous genus *Vanilla*) do occur across the remnants of Gondwana. The orchids in this subfamily have both genetic and structural features (such as loose pollen) that suggest that they split off from other orchid lines a long time ago, which makes sense, judging from their apparent ancient Gondwana connection.

Just as questions of timing of evolutionary events can be approached by looking at modern species, so too, questions about the history of past changes in a lineage can be answered by making the right kinds of comparisons among the modern members of that group. If, for example, we want to figure out how the warty hammer orchid came to be so darn strange, we might take advantage of the fact that there are some other living descendants of the defunct ancestral species from which the warty hammer orchid evolved. If we could identify the less heavily modified descendants of the ancestral species, we might get an idea about the kinds of traits exhibited by that extinct species, whose traits have been altered to greater or lesser degrees in the lines leading to its various evolutionary offspring. In other words, some living species may be more like an ancestral one. These modern organisms, with their evolutionarily "intermediate" traits, may help us reconstruct the sequence of changes that took place as an ancient proto-orchid gave rise to modern hammer orchids.

Comparisons of this sort require that we identify which species are related to one another by evolutionary descent. This task would be impossible if each living species was utterly unique. But because modern species resemble others to varying degrees, biologists have been able to place species into groups on the basis of shared oddities that set them apart from one another. We have already mentioned that every one of the twenty thousand to thirty thousand orchids has a column that is unlike that of any other flowering plant. Moreover, within the family Orchidaceae, different groups can be assembled and given different generic labels on the basis of their distinctive structural attributes. The warty hammer orchid shares its genus (*Drakaea*) with eight other species, all of which possess a very similar labellum with a three-dimensional female wasp decoy and a hinged rod substantially different from that of any other orchid species. The close similarity between the eight species of hammer orchids makes evolutionary sense if there was once an ancestral orchid with certain *Drakaea*-like features that gave rise to a branching lineage whose surviving twigs are the currently existing hammer orchids.

The most powerful means of establishing who is related to whom involves the use of various modern molecular techniques that were unknown to Darwin and the many other biologists who classified living things in the premolecular era. Today, however, we can compare species with respect to their genetic material. DNA is composed of an immensely long chain of four different building blocks called *nucleotides*. The four nucleotides differ with respect to one of their components, the bases, of which there are four types: adenine = A, guanine = G, cytosine = C, and thymine = T. Chains of nucleotides form unique sequences of immense variety with respect to their embedded bases (e.g., ... AATCGGCATTAA ... versus ... TATCGGCATTAA ... versus ... GGGGGGGGGAGG ... and so on and on). It is sometimes possible to compare the base sequence of DNA in its entirety from two or more species; alternatively, one can locate and examine a specific gene to see how similar its base sequence is for two or more carriers of the gene. This sort of comparative analysis can be used to determine which species are closely related to which other species on the principle that species with a very recent common ancestor will tend to share extremely similar DNAs, having inherited the same nucleotide sequence from that recent ancestor. The shorter the period of evolutionary separation, the less time for mutations (random changes) to occur that would alter the genetic material of the two populations, and therefore the greater the base sequence similarities between them.

Still another molecular means to the same end is to compare the similarities among species, not of their DNA, but of the proteins they produce. A protein molecule is composed of a chain of smaller molecules called amino acids, each linked to its neighbor in series. As it turns out, much of the functional information contained within DNA specifies a particular sequence of amino acids. Thus, the DNA base sequence AATCGGCATTAA contains sufficient information to code for a chain of four amino acids. The composition of this chain differs from the amino acid sequence encoded by most other base sequences of the same length, such as TATCGGCATTAA. One can therefore indirectly read the information in an organism's genes by deciphering the amino acid sequences of the proteins manufactured by that organism.

Both DNA and protein analyses have helped biologists check hypotheses on the evolutionary relationships among species, usually those that have been developed by comparing the major structural attributes of the species in question. For example, long before molecular biology came on line, humans and chimpanzees were considered to have become separate species relatively recently (in geological terms) on the basis of a host of shared skeletal similarities, such as the resemblance between human and chimp skulls. If it is true that these two species evolved from a common ancestor that lived not so long ago, then the DNA base sequences of the two species should be more similar than, say, human and orangutan base sequences, on the grounds that the lines leading to modern humans and orangutans supposedly separated farther back in time. Tests of this proposition have confirmed a remarkably high degree of similarity between human and chimp DNA. By some estimates, our base sequence matches that of the chimpanzee by almost 99 percent, a slightly higher figure than that for humans and orangutans and considerably higher than that for humans and most other primates, let alone other mammals.

Comparative DNA work has therefore established the close relatedness (that is, recent separation) of humans and chimpanzees. But molecular work can also reveal some decidedly counterintuitive connections, such as the fact that orchids belong to a group of plants, the order Asparagales, which includes asparagus, irises, daffodils, and agaves, none of which looks particularly orchid-like. Yet their substantial DNA similarities leave little doubt that this collection of plants constitutes a cluster of relatives. All these living species are part of a highly branched lineage that owes its existence to a common ancestor, which endowed all its descendants with a distinctive genetic heritage.

Yet, given enough time, common ancestry may not prevent considerable divergence in the attributes of a cluster of descendants. The warty hammer orchid's bizarre flowers are markedly different from the equally strange flowers of even its putative close relatives, the flying duck orchids, let alone those of more distant relatives, such as the sun orchids. Persons interested in adaptation ask how the hammer orchid's flower parts promote pollination; persons interested in history would like to know the answers to different kinds of questions about the antecedents of the warty hammer orchid's flower and the modifications of the ancestral flower that have resulted in this superbly odd species.

Darwin would have enjoyed doing the detective work needed to get at the history behind the warty hammer orchid and its bizarre flowers. However, he never made the acquaintance of this or any other hammer orchid, and so never had a chance to wonder about the origins of the warty labellum of this species. Darwin did visit Western Australia in 1836 during his round-the-world tour on the *Beagle*, but the ship remained in King George Sound on the south coast for only eight days in March. During this time, Darwin attended an aboriginal corroboree and collected some rocks, fossils, and insects but no plants. He seems not to have been terribly taken by the Western Australian countryside; his account of his research in the area is only two pages long, and his summary of the visit is not favorable, to put it mildly: "I do not remember, since leaving England, having passed a more dull, uninteresting time" (quoted in Patrick Armstrong, *Charles Darwin in Western Australia*, p. 39).

If Darwin had only come to Albany during the orchid season, his evaluation of Western Australia would have been very different. March is one of the least orchidaceous of months on the south coast of Western Australia. August and September are far better, a time when several species of hammer orchids are in flower. Had Darwin come in these months and had he found a warty hammer orchid in bloom, he would have been entranced by the plant, even more so if he had learned that the orchid's pollination depends on a single wasp species with an inclination to treat the orchid labellum as if it were a female wasp. Darwin would surely have wondered about the evolutionary predecessor of this extraordinarily odd flower petal.

No doubt he would have realized that because the vast majority of insect-pollinated plants, including most orchids, have flowers with more typical flower-like petals and sepals, it is reasonable to assume that somewhere back in time there was a more ordinary-looking orchid ancestor of the warty hammer orchid. Indeed, the subtribe (Drakaeinae) that contains the warty hammer orchid has been shown via molecular comparisons to be fairly closely related to two other subtribes, the sun orchids (Thelymitrinae) and the donkey orchids (Diuridinae), both of which have more traditional-looking flowers. If these groups have flowers with similarities to those of an ancestor of the hammer orchids, then we need to figure out how to get from flowers with six "petals" of more or less the same size and shape to an orchid endowed with one exceptionally modified petal among its set of six.

One way to produce a tentative hypothesis on how something like a sun orchid could eventually give rise to a hammer orchid is to examine the floral diversity exhibited by members of yet another orchid group, the Caladeiinae, which contains the especially species-rich genus *Caladenia*. As noted already, many *Caladenia* orchids have a spidery appearance, thanks to their almost absurdly elongate sepals and petals, which gives them their common name, spider orchids. But unlike the hammer orchids, all of which are very similar in appearance, *Caladenia* flowers come in a wide variety of shapes and sizes, some of which resemble the flowers of hammer orchids in that the labellum has evolved into a female decoy. The zebra orchid, for example, is a case in point (see Figure 3.1). By comparing the flowers of living caladenias, we might at least be able to generate a hypothesis about the changes that have taken place over evolutionary time, changes that involved the gradual transformation of the flowers in an ancestral lineage.

One possible sequence of historical events begins with a relatively ancient species adorned with flowers composed of ordinary-looking petals. The ancestor then split into two descendant species, each of which in turn split again into newer, more recent orchids, the process repeating itself again and again, eventually yielding some species with one petal endowed with some of the attributes of a female decoy. As I say, we can get some idea about what this range of ancestral species might have looked like by comparing living species, not because some of today's species are "missing links" of some sort but because certain plants may have retained an only modestly modified ancestral attribute that happened to work well in their environment right up to the present.

4.1

Donkey orchids (genus *Diuris*, left) and sun orchids (genus *Thelymitra*, right) have flowers that are more familiar, at least superficially, than those of hammer orchids and other genera with elaborate female decoys.

When one surveys the caladenias of Australia, it doesn't take long to find species whose lip petals do have characteristics that appear to be somewhere between an ordinary, unexceptional flower petal and the very much out-of-the-ordinary female decoy petal of the hammer orchid. For example, many spider orchids have a more or less normal-looking labellar petal that is, however, endowed with several rows of small, inconspicuous bumps, or calli, which, as you may recall, are the glandular, possibly sex-pheromone-producing structures found on the "head" of the hammer orchids' decoy. The small pale calli of certain spider orchids (and other flowering plants as well) are part of the secretory apparatus that the plants use to generate scents to attract pollinators, sometimes alerting insects to the presence of a rewarding nectar source and sometimes, less charitably, exploiting the sex drive of certain male insects. Some spider orchids also have other, thickened hair-like glands on their sepals and sometimes petals that produce attracting odors. These volatiles may draw the pollinator in close enough to permit it to see the calli on the labellum, which orient the wasp as it attempts to copulate with the labellar petal or a portion thereof.

Considerable variation in calli size, color, and distribution exists among living species of *Caladenia*. The big, beautiful, white spider orchid *Caladenia longicauda* has pale purple calli on its gorgeous white

4.2

A comparison of several species of *Caladenia* orchids suggests a
hypothesis for the evolution of elaborate female decoys in orchids.
(opposite, top) Decoys may have originated in the form of widely
separated rows of small pale calli on a lip petal, as seen in this species,
Caladenia vulgata. (opposite, middle) Subsequently, the scent-
producing calli may have become larger, darker, and more closely
spaced, as in *Caladenia wanosa.* (below) In both the bee orchid,
Caladenia discoidea, and (opposite, bottom) the leaping spider
orchid, *Caladenia macrostylis,* the lump of calli lying in the center
of the labellum provides a somewhat more convincing, but still very
generalized, female decoy to deceive male wasp pollinators.

Caladenia discoidea

Caladenia vulgata

Caladenia wanosa

Caladenia macrostylis

labellum, but they are widely spaced and do not form anything like an obvious female wasp decoy, which makes sense because it is pollinated by bees. Other species, such as the Kalbarri spider orchid, *Caladenia wanosa*, have larger, darker, more closely packed calli on the labellum, creating a more conspicuous array than in *Caladenia longicauda*, but the calli do not form a mimetic insect body of any sort. Even so, the combination of visual and olfactory stimuli is enough to attract male thynnine wasps. For others, however, the calli have coalesced, forming a conspicuous dark lump that lies seductively in the center of the labellum, as in the bee orchid, *Caladenia discoidea*, and another member of the genus, *Caladenia macrostylis*, the leaping spider orchid. If the petal were to become further restructured, so as to feature a condensed mass of calli at the head of a curled, tubular petal, then we would have a female decoy not unlike that of the modern hammer orchids.

As for the elongate, flexible stalk that holds the hammer orchid's decoy at arm's length from the column, here, too, one can find less extraordinary analogues among the spider orchids. In many of these species, the labellum is attached to the rest of the flower by a short but flexible curl of tissue. As a result, the labellum can easily be made to move when a pollinator interacts with it. These movements facilitate contact between the insect visitor and the pollen-bearing column. In some cases, the system works in a manner highly reminiscent of that of the hammer orchid. Thus, the so-called lazy spider orchid, *Caladenia multiclavia*, has a magnificent red-striped labellum with a coalesced lump of calli on its anterior portion. The labellum is attached to the flower by a short, flexible stalk. Should a wasp land on the calli, the labellum will tip over, putting the wasp upside down and bringing the insect in contact with the column in much the same way that the warty hammer orchid manipulates its pollinator.

The point is that by comparing living species of *Caladenia*, we can develop ideas about possible intermediates between a flower petal that

4.3
The lazy spider orchid, *Caladenia multiclavia*, has evolved a pollination system with similarities to that of the warty hammer orchid, including a calli-packed female decoy on a labellum that sits on a flexible, hinged stalk.

lacks even a hint of female decoy to a flower petal that is all female decoy. The current existence of these diverse forms tells us that it is not implausible to propose that a similar range of flower types existed in the past, providing a way to get from trait A to trait F in an evolving lineage of hammer orchids without requiring magisterial leaps or wild jumps.

Although Darwin did not employ this approach in great detail when dealing with orchids, he explicitly applied this line of reasoning in his studies of sundews and other carnivorous plants. His goal was to show how the animalian traits of sundews could have evolved from those of perfectly ordinary plants, most of which manage quite nicely on a diet of sunlight and carbon dioxide without any fly or bug supplements. To account for the evolution of insect-eating plants capable of attracting, capturing, maneuvering, digesting, and absorbing nutrients from their prey, Darwin showed that all of these carnivorous abilities rested on attributes exhibited by other plants with a completely traditional lifestyle. So, for example, the sticky tentacles of a sundew are much larger and more elaborate than the glandular devices found on the leaves of many non-insectivorous plants, but sundew tentacles are not a quantum leap into another universe. In fact, some rather ordinary plants more or less accidentally entrap unlucky insects thanks to very small amounts of stickum exuded by certain cells in their leaves.

4.4

A Western Australian sundew that has trapped a small fly with its sticky tentacles.

A case in point: the sticky purple geranium with the wonderful Latin name *Geranium viscosissimum*. When small insects land on the geranium's leaves with their viscous coating, the bugs may be repelled and immediately fly off, or they may become glued to a leaf, where they die a slow death. In either case, the plant benefits by thwarting herbivores that would otherwise eat its leaves. From this finding, Darwin deduced that the sticky secretions of a distant ancestor of the sundew probably

The History in Evolution

initially served a defensive function, not a nutrient-acquisition function. But once a trapped insect had gone to its reward, the substances it contained might ooze onto the leaf, where they could be absorbed by the plant's cells. And the sticky purple geranium is indeed able to take in proteins from its deceased insect "prey," thereby acquiring some useful chemicals as a result. This geranium and other similar species told Darwin that the ancestor of today's marvelous sundews might well have been a pretty standard plant whose leaves happened to secrete small amounts of a sticky material. Herbivorous insects may have generally avoided these plants but were sometimes unlucky enough to become captured.

The probability of prey capture might increase if the defensive secretory hairs generated larger amounts of glue. There are living plants of this sort, including the attractive blue-flowered *Plumbago*, which has quite large glandular hairs arrayed about the bases of its tubular flowers. The sticky hairs deter insects that would otherwise steal nectar from the plant by gnawing their way into the corolla, short-circuiting the pollination system of *Plumbago*, which requires that a long-tongued pollinator insert its proboscis down the tubular flower.

Thus, a plant ancestral to today's sundews might well have had secretory hairs of some size, which, with only modest elaboration, could enable this species to become a proto-carnivore with the ability to take advantage of the nutrients in its incidental captives. Interestingly, the family to which *Plumbago* belongs contains a number of other glandular genera, and the family as a whole is not too distantly related to the sundew family. Conceivably, a *Plumbago*-like ancestral species could have given rise to both the Plumbaginaceae and the Droseraceae, a hypothesis that is entirely testable in principle, as I shall illustrate with the case of the Venus flytrap, the most famous carnivorous plant and arguably the one with the most complex fly-catching apparatus.

4.5

The evolutionary relationships of the sundews (*Drosera*) and related families. Hypothetical extinct ancestors are given letter labels pasted on the tree of life diagram next to a description of their probable mode of carnivory.

The adaptive value of the flytrap is reasonably clear, but just how such a wonderful device originated is not immediately obvious. One possible solution to the problem of origin requires that the Venus flytrap be in effect an elaborated sundew. This hypothesis produces the prediction that the Venus flytrap and the sundews should share genetic information as a result of having descended from the same, now extinct, ancestor. In fact, DNA data now demonstrate that the Venus flytrap and the sundews are indeed close relatives despite their very different appearance and distinct methods of carnivory. In recognition of their shared ancestry, the two are placed in the same family (Droseraceae) but different genera (*Dionaea* and *Drosera*). The fact that all the many members of the genus *Drosera* as well as the one species in a related family, the Drosophyllaceae, have flypaper traps composed of sticky hairs suggests that the ancestor of both families possessed a flypaper trap (extinct ancestral species A in Figure 4.5). Thus, the Venus flytrap probably originated from an ancestor with flypaper leaves. With the passage of time, a

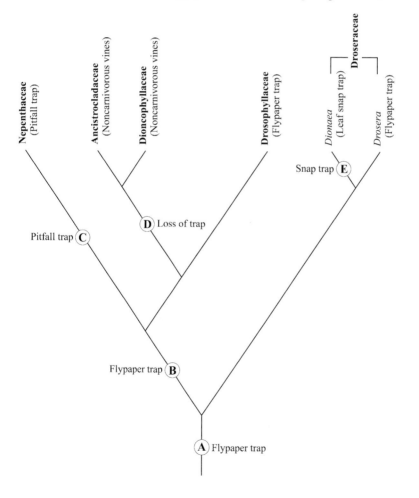

carnivorous plant with flypaper leaves might have first added slow leaf movements of the sort now seen in sundew tentacles (and remember that the whole leaf of some sundews can bend inward very slowly in response to the capture of prey). Branching off from the sundews came a line that evolved differently, resulting in a now extinct species able to move its lobed leaves relatively rapidly. This sort of plant produced descendants, one of which is today's Venus flytrap, a greatly modified survivor. The Venus flytrap has secretory glands that are not associated with large rod-like hairs, and it uses its leaves as a snap trap, changes that must have taken place *after* effective stickum traps had evolved in its evolutionary lineage.

Another snap trap plant exists, the waterwheel (in the genus *Aldrovanda*). Although a thin aquatic plant that looks rather like the common waterweed (*Elodea*), its leaves are actually nifty little traps capable of snapping up small aquatic crustaceans and the like. Very recent genetic analyses have demonstrated beyond much doubt that the Venus flytrap and the waterwheel are also close relatives, which makes them both "cousins" to *Drosera* sundews. This tells us that snap traps have only evolved once somewhere along the line leading to the Venus flytrap and its aquatic sister, the waterwheel.

Further examination of the evolutionary relationships of the five families of plants shown in Figure 4.5 offers evidence about the history of assorted other changes that have occurred during the evolution of these families. The fact that all the modern members of the Ancistrocladaceae, a family that includes various tropical vines, and all but one species of the Dioncophyllaceae, another tropical plant family of lianas, are *not* carnivores is consistent with the loss of carnivory in an extinct species (D) that gave rise to both families. If this scenario is correct, the fact that *Triphophyllum peltatum* (a member of the Dioncophyllaceae) has glandular hairs and is potentially carnivorous during one growth phase can be interpreted as the reacquisition of carnivory by this one species of liana from a non-meat-eating ancestor of intermediate age.

If true, what we have here is a case of convergent evolution, the evolution of a functionally similar trait, sticky hairs capable of trapping insects, originating from two different starting points in two different lineages of organisms. Convergent evolution has happened repeatedly during the history of life. For example, although the five flypaper trap species in the genus *Byblis* have a host of sundew-like sticky hairs on their stems, they are no longer believed to be relatives to the mucilaginous sundews. Molecular data tell us that these plants are instead quite closely related to the Solanaceae, the plant family that contains the tomato and potato. Evidently, *Byblis* evolved its fly-catching apparatus completely independently of *Drosera* and *Drosophyllum*, despite the superficial similarities in the sticky hairs of these different plants.

One other family, the Nepenthaceae, *is* included in the cluster that contains the sundews because of its molecular similarities with the sundews and its close relatives. The species in the Nepenthaceae possess a remarkable carnivorous trap that has no resemblance to a sundew's leaf. Instead of a flypaper leaf, species of *Nepenthes* come equipped with leaves endowed with a bizarre tip that has been greatly extended and shaped into a tubular trap. Insects that land on the waxy upper lip of the trap, which looks rather like the pitcher of the carnivorous pitcher plant, sometimes slip into the vessel and are drowned and digested in the fluids contained within. One can hardly imagine that such a distinctive device could have evolved from a flat sticky leaf, but the evolutionary relationships among living plants based on molecular comparisons tell us that this is exactly what must have happened. The basal ancestors of *Nepenthes* (extinct species A and B in Figure 4.5) presumably had sticky leaves because so many of their modern descendants do as well. If so, the pitfall leaf trap of the Nepenthaceae was an evolutionary innovation derived from flypaper

4.6

The purple-flowered Western Australian *Byblis* (above) is not closely related to the sundews, but the weird pitfall trap plant, *Nepenthes* (right), does share a fairly recent common ancestor with the sundews.

leaves just as the snap traps of the Venus flytrap and the waterwheel are also highly modified derivatives of a flypaper trap. Although the traps of *Nepenthes* do not have sticky hairs, they are endowed with a host of digestive glands, which presumably evolved from glands that do double duty in sundews by first trapping and then digesting insect prey. In any event, the genetic similarities among certain of the carnivorous plants tell us that the amazing leaves of *Nepenthes* must have evolved directly from the slightly less amazing leaves of a flypaper ancestor.

Let's now return to the hammer orchid's evolutionary history to generate hypotheses about the reasons why they may have gradually evolved a female decoy instead of relying on the more typical flowers of their putative distant ancestor. Perhaps at some point in its evolutionary lineage, an orchid had proto-calli that contributed attractive odors to the blend of floral scents released by the plant to advertize the nectar reward waiting for an insect. If, however, one of these volatile floral compounds in an ancestral orchid happened to be modified by a mutation so that it had even a slight resemblance to an insect sex pheromone, the individual(s) with the mutant gene and different odor could have enjoyed a pollination advantage. These plants might have attracted some pollinators that were hunting for nectar and some others that were seeking mates. If the double-barreled pollination system improved the plant's odds of transferring and receiving pollen, the mixed blend attractant could spread through the species, setting the stage for additional modifications, each of which increased in frequency or disappeared over time depending only on whether the novel scent mixture happened to lead to slightly higher or slightly lower reproductive success.

An alternative scenario for the origin and evolution of pseudo-sex pheromones by orchids involves the waxes these plants often produce for protection and waterproofing. As it happens, the cuticular hydrocarbons found in the waxy coats of at least some orchids are similar to those found in the cuticle ("skin" or exoskeleton) of some insects. If a mutant orchid produced waxy substances that happened to attract sexually motivated male bees, this plant might have set in motion the process of change by producing more offspring than others of its species. Over time, ever better chemical mimics could spread through the orchid species if these pheromone-mimicking individuals had access to a larger or more efficient pool of pollinators.

This reconstruction of historical events has received support from chemical studies of the European bee-mimicking orchid *Ophrys sphegodes* and its pollinator, a bee called *Andrena nigroaenea*. When biochemists extracted the hydrocarbons from the cuticle of the female bee, they secured a complex blend of compounds, at least fifteen of which elicit a response from olfactory receptors in the antennae of male bees. All fifteen of these hydrocarbons can also be dissolved off the bodies of the decoy labellum of the *Ophrys* orchid (this is a genus with many wonderful examples of lip petals that look remarkably like female bees). Not only that, but the relative amounts of the various compounds are essentially the same for the orchid decoy and the real McCoy. If one applies compounds extracted from either orchid decoys or female bees to the body of a dead female bee from which all scents have been removed, the probability that the dummy will be approached and pounced on by male bees is the same. Thus, the males do not discriminate between the orchid's scent and that of a living female bee.

The suggestion has been made for *Ophrys*, at least, that the waxy coating of petals, which originally evolved for its waterproofing value, has become modified because the scents incidentally happened to attract pollinators. One of the benefits of this change would have been a reduced need to manufacture other floral scent compounds, which require energy to make, an investment that could be saved if plant waxes were used instead. Today's *Ophrys* emit almost no standard floral scents. Instead, they smell like female bees.

If, by any means, floral compounds could be replaced by an increasingly effective false sex pheromone, individual orchids could have secured several important benefits in addition to a reduced need to produce standard floral volatiles. The sexually deceptive plants could also stop investing in unneeded nectar production, thereby saving energy for other reproductive purposes. In fact, about a third of all orchids offer their pollinators no food reward of any sort. Many of these species are the sexually deceptive marvels that we have focused on, although other forms of orchid trickery exist, including some nectarless species whose flowers are endowed with small bumps and protuberances apparently designed to look like aphids. The pseudo-aphids may trick female hover flies into landing on the flowers because these flies like to lay their eggs on aphid-infested plants to provide their predatory larvae with easy access to aphid prey. After landing on an aphid-mimicking orchid in search of a productive oviposition site, the fly may deposit some eggs, dooming these offspring to starvation when the eggs hatch. The orchid could care less, of course, as long as the tricked fly inadvertently picks up or deposits pollen in the course of its egg-laying visit.

Yet another form of deception consists of the mimicry of other nectar-producing plants by sugar-free orchids. Unlike those honest orchids that provide sugary fluids in return for the services of certain insects, these dishonest mimics merely look as if they have nectar on tap, even though they do not. By encouraging a foraging insect to search the flower carefully but fruitlessly for a nonexistent reward, the deceptive orchid may succeed in having some visitors come in contact with its pollinia before departing. If this hypothesis is correct, the addition of an experimental reward to a deceptive orchid should actually *lower* the probability of pollinia removal, which is exactly what happened when Ann Smithson and

Luc Gigord added a sugar solution to the naturally rewardless flowers of the European orchid *Barlia robertiana*. These experimenters found that bees failed to energetically probe flowers that had received added droplets of sugar solution and therefore often failed to come in contact with the pollen masses. In contrast, bees that foraged on natural rewardless flowers kept pushing into the flower in search of nectar that was not there, and in so doing, they regularly touched the orchid anther and came away with the pollinia.

Perhaps the same trick works for the Western Australian pink enamel orchid, *Elythranthera emarginata*, a species with flowers that are big, lovely, shiny, showy, and easy to locate, all typical attributes of the many "honest" flowers that really do contain food for pollinators. But enamel orchids neglect to sweeten their flowers with nectar, something pollinators learn only after they have responded to the flashy floral come-on of the orchid.

The Australian donkey orchids in the genus *Diuris* cheat in a similar way. Their large flowers may look like those of nectar-offering pea plants that grow nearby, but instead of a food reward, the typical donkey orchid

4·7
Deceptive pink enamel orchids, *Elythranthera emarginata*, are easy for pollinators to locate but do not offer a food reward for their visitors.

4.8
Donkey orchids use floral mimicry to deceive insects, like this introduced honey bee, into visiting their flowers (top), which resemble the flowers of reward-supplying plants, like this member of the Papilionaceae (bottom).

offers nothing edible at all. When small native bees searching for pollen land on the conspicuous flowers of the orchid, having learned from encounters with nearby pea plants that flowers with this color pattern and shape are worth inspecting, the flower may stick its pollen on the bee for transport to another plant. The bee gets nothing in return for its services.

In eastern Australia, the black-tongue spider orchid *Caladenia congesta* plays yet another kind of trick on pollen-searching bees. This orchid has a labellum that consists largely of a mass of black calli, which projects out rudely from the flower. One could easily assume, as I did initially, that here is another female decoy designed to arouse sexually excitable male wasps or bees. And it is true that certain bees do fly to and grasp the black lip petal, but the current view is that they do so in an effort to shake pollen from what appears to them to be an anther. Several Australian plants do have blackish porous anthers that require rapid shaking to dislodge the pollen within, which can

4.9

In the black tongue orchid, *Caladenia congesta*, the calli have evolved into a dark cylindrical mass that is believed to deceive pollinators that have had experience with the rewarding, pollen-rich anthers of certain other plants.

then be harvested by the bee. Apparently, *Caladenia congesta* mimics these so-called buzz-pollinated plants to lure bees to the labellum, which the bees vibrate in a futile attempt to shake nonexistent pollen free from the structure. In so doing, the hard-working bees may get plastered with pollinia from the orchid column.

Food reward deception is also practiced by a South African orchid, *Disa pulchra*, which closely resembles the iris *Watsonia lepida*. Both plants produce a spike with twenty or so showy pink flowers so similar in appearance that people often get the two mixed up. So does a horse fly with a long tongue, which it uses to extract nectar from the rewarding flowers of the iris. When the fly mistakenly lands on the orchid and sticks its proboscis into a flower, it gets nothing except a pollinarium stuck to the base of its proboscis. After probing just one or two equally empty flowers on the orchid's inflorescence, the fly is on its way, perhaps to be fooled again, in which case the insect may deposit pollen collected at the first specimen onto another orchid's stigma. Crosses of this sort produce far more viable seeds in *Disa pulchra* than cases in which the orchid reproduces by "selfing," that is, by using its own pollen to pollinate its own flowers.

Indeed, one reason a multiflowered orchid might reduce the rewards offered to pollinators is to encourage a visitor to leave quickly, rather than hang around going from one flower to another and another on the plant's flower stalk. This hypothesis provides an explanation for a reduction in nectar production that differs from the "make the pollinator work to pick up the pollinia" idea that we discussed earlier. The more rapidly the potential pollinator leaves an orchid, carrying a pollen package from the plant, the less likely it is to keep interacting with that orchid's other flowers. Such an outcome reduces the risk that the plant will get self-pollinated, with all the problems attendant on inbreeding (for many species). When a pollinator interacts with a nonrewarding orchid flower, it has every motivation to leave the plant quickly after discovering that it has been cheated.

Experimental evidence in support of this expectation comes from studies by Steven Johnson and his colleagues. In one project, they added nectar to nonrewarding deceptive *Orchis* flowers and found that female bumblebees stayed longer on plants with the artificial reward than on those that were their usual deceptive selves. In nature, therefore, bees may help the unrewarding deceptive orchid outbreed.

In a similar project with another bumblebee-pollinated orchid, *Anacamptis morio*, Johnson and company added nectar to the spurs of this orchid's flowers, which encouraged the bees to stay longer on each plant and visit more flowers. As a result, bees regularly stayed around longer than the fifteen seconds or so required for the removed pollen package attached to the bee's proboscis to bend over in the position that makes pollination possible. (You may recall that Darwin observed the time-delayed bending of the stalked pollen-bearing device in *Anacamptis pyramidalis*—Figure 1.6—which he proposed was an adaptation for outcrossing.)

The cross-pollination enhancement hypothesis also appears to apply to the sexually deceptive, rewardless Australian orchid *Caladenia ten-*

tactulata. Rod Peakall and Andrew Beattie reached this conclusion after injecting the pollinia of selected orchids with variously colored dyes and then later surveying the surrounding population of orchids for marked pollinia that had been transferred to a recipient plant. They found sixty-six distinctively dyed pollinia that wasps had taken from one orchid and applied to another. The mean distance that the pollinia had been carried was 17 meters; therefore, the plants were obviously not self-pollinating, nor were they breeding with close relatives clustered around them. Outbreeding is advantageous in this species; offspring of outbred crosses develop more rapidly than inbred offspring.

As these cases illustrate, a "cheating" orchid not only avoids the costs of manufacturing nectar, it is also likely to acquire a specialist pollinator good at distributing its pollen to appropriate targets some distance away. Only certain bees target the black tongue orchid's false stamen, and only the flies that are attracted to *Watsonia* irises will be lured to *Disa pulchra*. Likewise, only the male thynnine wasps that approach the sex pheromone produced by their females will be attracted by hammer orchids able to mimic those special odors.

Thus, the gradual process of decreasing nectar production and improving pheromonal mimicry could yield any of several benefits at each small evolutionary step of the way. All that is required is that some small changes enabled some individuals in the past to be ever so slightly better than their fellows at deceiving their pollinators. Once one mutant gene with a small beneficial developmental effect had spread through a population, the stage was set for the incorporation of an additional reproduction-enhancing mutation. As more and more changes were layered on past ones, the cumulative effect on the development and appearance of the species could have been great indeed.

Now it is true that large evolutionary changes need not always be gradual. In recent years, developmental biologists have discovered that some genes act as switches, controlling the activity of many others. This class of genes therefore can have a major influence on how organisms develop. In turn, a mutation in one of these regulatory genes has the potential to generate large and dramatic changes in body building. If any of these changes happened to raise, rather than lower, reproductive success, a single mutation could lead to something of an evolutionary jump instead of a baby step. So, for example, William McGinnis and his co-workers at the University of California, San Diego, believe that the distinctive six-legged body plan associated with insects may have arisen abruptly from a much different body design in which a segmented body had paired limbs on each and every segment, not just three. That such a big change was possible in the past is supported by experimental studies of the effects of mutations in a group of regulatory genes, known as *Hox* genes, on leg development in modern brine shrimp. Some mutant *Hox* genes "turn off" leg development on all the abdominal segments of the shrimp, illustrating how a similar kind of regulatory change in an ancestor of insects could have restricted leg development to the thoracic region of that ancestor's body. If such a creature with a reduced number of legs gained some reproductive advantage from the mutation (perhaps an energetic savings from having to build fewer legs), then the mutant animal could have survived its dramatic reconfiguration and even have flourished. Such a creature was probably the ancestral animal from which all of today's millions of insect species are descended.

Although it is also possible that large developmental mutations could have contributed to the evolution of some aspects of orchid floral structure, we do not need to invoke massive, one-step changes to account for the great diversity of flower forms in the group. In all species, small changes can spread themselves if they raise individual reproductive success even slightly. The process will eliminate less effective alternatives and can lead gradually to the great modification of entire species as a whole series of changes replace their antecedents. The end result could be the production of the marvels that amaze and delight us all.

One such marvel is the approximately foot-long labellum of the star orchid (*Angraecum sesquipedale*) of Madagascar. The base of this thin tube-like floral spur provides a nectar reward—but only for a pollinator capable of reaching the fluid at the end of the tube. Darwin knew of the orchid, and he bravely predicted that a hawkmoth must exist in Madagascar whose proboscis was about a foot long, on the grounds that only such an insect could reach far enough into the spur to extract the nectar within.

Floral spurs, or nectaries as they are also known, occur in many orchids, and in at least one well-studied South African case, the length of the spur is indeed matched by the length of the pollinator's proboscis. The orchid in question, *Disa draconis*, has spurs that are 30 to 80 millimeters long (roughly one to three inches). In areas where most of the orchids have short spurs, the plants are visited by a fly with a proboscis in the 20 to 35 millimeter range; in other places where long-spurred plants are common, the orchids are pollinated by a different fly species with a proboscis in the 50 to 60 millimeter range. When a fly of either species inserts its proboscis deep into the spur, the orchid pollen masses attach themselves to the pollinator's elongate tongue. The attachment point depends on the length of the spur. After acquiring a pollen packet, the fly may visit another orchid of the same spur type, depositing the pollen onto the flower's stigma, and in so doing, inadvertently help the plant reproduce.

The importance of the close correlation between the length of floral spur and that of the fly's tongue was shown in an ingenious experiment conducted by S. D. Johnson and K. E. Steiner. They effectively shortened a sample of long-spurred orchids by tying the tube off partway down its length. These flowers were still visited by the long-tongued flies, but only a third of the pollinators departed with pollinaria attached to their proboscis (instead of over half, in the case of those control specimens whose long floral spurs were not altered). Moreover, pollen-carrying flies placed pollen on the stigma of only a sixth of the experimentally manipulated orchids, whereas successful fertilization occurred in about half of the unchanged control group.

This study demonstrates how important it is for an orchid to have the right match between the length of its spur and a pollinator's proboscis. The experiment was done in 1997, a hundred years after Darwin. No hawkmoth with a foot-long proboscis was known when Darwin made his pronouncement, and many entomologists at the time openly doubted that one with such an absurdly long tongue would ever be found. However, around the turn of the century, unfortunately after Darwin's demise, the pollinator of the orchid was discovered, and, as the father of natural selection had predicted, it was a hawkmoth with a twelve-inch tongue. Appropriately enough, the moth was named *Xanthopan morganii praedicta*.

Because most hawkmoths have a much shorter proboscis and most orchids have much shorter floral spurs (if they have such a device at all), it stands to reason that both *Xanthopan morganii praedicta* and the star orchid evolved from ancestors with more ordinary tongues and spurs, respectively. Perhaps the moth and the orchid evolved together, with the moths slowly acquiring ever longer tongues in concert with orchids whose spurs became ever deeper. These changes could have occurred if individual moths that inherited longer-than-average tongues were able to secure nectar unavailable to their competitors. With the extra energy at their disposal, these long-tongued moths would have had the wherewithal to outsurvive and outreproduce others in their species that had somewhat shorter tongues. On the orchid front, individuals with deeper-than-average nectaries may have been able to attract more reliable specialist pollinators that would visit only plants like them, ensuring that their pollen was used efficiently to form more seeds than those plants with shorter spurs and easier-to-remove nectar.

Here, then, we have a possible case in which changes in one member of a pair of interacting species favor reciprocal changes in the other, which sets the stage for the next round of selection of the same type. According to this scenario, as the moths gradually evolved longer tongues, the orchids gradually evolved longer nectary spurs, producing a matched set of changes in the two organisms.

But there is another hypothesis for the evolution of the extraordinarily long floral spurs of the star orchid. According to this evolutionary scenario, the hawkmoth gradually evolved a longer and longer tongue, not to reach into a deeper and deeper floral spur, but to be able to feed while hovering well away from a more ordinary nectar-producing flower. Hawkmoths with extra long tongues can take up nectar without jamming their heads into tubular flowers. By keeping their distance and shifting position erratically, a moth can make itself less vulnerable to attack by a predatory mantis or spider lurking on or near the flower. In addition, the moth can also react more effectively to fast-flying bats zooming in for the kill. Thus, according to the biologist Lutz Wasserthal, selection by predators might have been responsible for the spread of genetic mutations associated with an increase in the length of the moth's tongue, which helped the insect avoid becoming a sitting duck while foraging.

Wasserthal argues that once moths with longer-than-average tongues existed, they could have successfully visited orchids that happened to have a nectar source at the end of a longer-than-average tube or spur. If these orchids offered sufficient rewards, they might attract moths willing to plunge their proboscis deep within the flower, even if this forced the insect to hover in a fixed position to secure the valuable nectar supply within. According to this scenario, at least initially, the orchids and moths were not locked in a reciprocal, coevolutionary embrace. Instead, the elongate orchid nectary evolved *after* the moth's tongue had become unusually long in response to predator pressure. Although debate on this matter continues, both of these historical hypotheses involve the layering of modifications on earlier changes in the manner envisioned by Darwin, leading eventually to most unusual attributes, namely, foot-long moth tongues and foot-long floral spurs.

Of course, some people take quite a different approach when analyzing the question of biological origins. They look at the extraordinarily complex attributes of living things, such as the intricate design of orchid flowers and the equally special features of orchid pollinators, and see evidence that God, not natural selection, has shaped the attributes of living things. Their assertion that God did it takes two forms, one unsophisticated but honest and the other slightly more sophisticated but essentially devious. In the first category, some fundamentalist Protestants make no effort to conceal the religious motivation for their creationism and matching antagonism to "evilution." Thus, Geoff Chapman, honorary secretary of the Creation Resources Trust in England, writes of the intricate design and complexity of the European orchid flowers that exploit bees as their pollinators and concludes that they "must have been created and designed to operate this way from the very beginning, and bear abundant witness to the design and power of God, the Creator." For Chapman, the clinching argument is that "if an orchid needed to look like a bee or other insect in order to attract a pollinator, then until it bore a significant resemblance, the insect would not be interested" ("Orchids: A Witness to the Creator").

This claim is based on the old argument that if one were to remove even one part of a complex machine, it would not work. (But note that small glitches in a computer may reduce the efficiency of the machine without necessarily eliminating its function altogether.) The analogy between hypersensitive man-made machine and complex living machines is invoked to suggest that natural processes could not produce a multi-component organism via a series of changes from a simple predecessor to an elaborate current living thing via a long series of operational intermediates. This argument is so old that Darwin had a chance to deal with it, which he did. For example, he showed that the extraordinary features of carnivorous sundews were present in simpler form in some noncarnivorous plants, just as the masterfully complex eyes of vertebrates are elaborations of less complex, yet perfectly functional light sensors in other animals. Because these simpler leaves and simpler eyes do useful things for their owners now, they surely could have worked in the past and thus been available for gradual modification over time in some lineages, leading eventually to the marvelous leaves of sundews and the glorious camera eyes of red-tailed hawks and the like.

With respect to the complex flowers of orchids with their integrated working parts, we have seen that many Australian orchids have attractant calli that lure male insects to the orchids even though the calli bear no "significant resemblance" to a female insect of any sort (see Figure 4.2). The same is true for many other orchids. Indeed, the petals and sepals of the South American *Trigonidium obtusum* have not the faintest resemblance to a female bee, but even so, male bees fly to these rather ordinary-looking petals and sepals and attempt to copulate with them. As they try their best, some males slip on the waxy surface of the petal and fall into the flower tube; needless to say, they do not get out until they find a passage that takes them right past the stigma and anther. Both *Trigonidium obtusum* and species like the Kalbarri spider orchid *Caladenia wanosa* tell us that a perfect decoy is most definitely *not* required right from the start to do the job of deceiving male wasps into visiting certain orchids. Instead, all that is needed at the outset is a flower petal that does an ever so slightly better job of attracting pollinators than whatever the standard device is at the time for a given species. Plants with slightly modified flowers that attracted a few more effective pollinators were likely to produce more seeds and leave more descendants, which could have inherited their parent's modified flower.

But, as I say, at least Chapman and many other religiously motivated persons do not try to hide the fact that what they are up to is not scientific. Having convinced themselves that God is responsible for all, they simply do not find a naturalistic explanation for complex orchid flowers plausible; nor do they seek one. And they make no bones about it. In the last analysis, their "counter" to evolutionary analysis is that they are sure it is wrong because they, personally, cannot imagine how such and such a trait could evolve gradually via standard evolutionary processes. This view is not persuasive to most noncreationists because it suffers from the defects of what Richard Dawkins has called the "argument from personal incredulity." As Dawkins explains in his book, *The Blind Watchmaker*, such an argument can always be rejected if someone less incredulous is able and willing to think of a plausible explanation for the very phenomenon that others say cannot be explained. Although creationists may not be able to imagine how orchid pollination systems could have evolved, botanists and evolutionary biologists have not been similarly befuddled, as we have seen.

It is revealing that two other adaptive complexities often cited by creationists as inexplicable—deep oceanic diving by whales and flight by birds—have both been thoroughly explicated by evolutionists. Ample fossil evidence has permitted the reconstruction of history showing that the oldest whale-like creatures, the pakicetids, lived about 50 million years ago and were mammals about the size of a fox or wolf. We know that pakicetids have to be the ancestors of modern whales because their tympanic bone, a portion of the middle ear, had a distinctive fold on one side in the shape of an S. All modern whales have this unique anatomical peculiarity; no other mammals possess the sigmoid process. Therefore, pakicetids were whales.

At one time, when only the skulls of pakicetids had been discovered, it was thought that these animals might have been amphibious, alligator-like predators. But when the leg bones of *Pakicetus* and a related species were finally found, they proved to be long and slender, the kind found in legs designed for running on land. Furthermore, the ankle bones of these creatures closely resembled those of the so-called even-toed ungulates. whose modern-day representatives include the cow and the hippopotamus. In other words, the pakicetids had the ear bones of seagoing whales and the leg bones of land-based mammals.

If there really was an evolutionary transition from a terrestrial ungulate to modern whales, at some time there must have been transitional forms that were amphibious. One can imagine roughly what such a creature might have been like given the fact that there are amphibious ungulates among the living mammals, notably the hippopotamus, which some paleontologists believe to be the whales' closest living relative. In the search for an extinct terrestrial-aquatic intermediate, biologists have found reasonably complete skeletons of a close relative of the pakicetids, an extinct mammal now called *Ambulocetus*, which lived after the first earth-bound whales had appeared. The skeletons of *Ambulocetus* have a host of features intermediate between that of pakicetids and modern, highly specialized whales. For starters, *Ambulocetus* had front and hind legs, big feet, and a long tail, not flippers or a tail fluke. This whale probably looked and behaved like an alligator or crocodile, sharing with these animals the capacity for an amphibious lifestyle in coastal habitats near shallow seas.

The whales that followed *Ambulocetus* became more and more independent of land and more and more adapted to life far out in the ocean. Their hind limbs became progressively smaller over time and are now only represented by tiny vestigial remnants in some living whales. Their front limbs became encased in flippers. The tail fluke emerged in conjunction with a shift from paddling to swimming via undulations of the vertebral column. Although today's sperm whale and bottlenose dolphin look nothing like a terrestrial, hairy, four-legged, hoofed mammal, the fossil record tells us that they are the descendants of just such an animal from which was derived a lineage containing more and more whale-like forms, as Darwinian theory predicts. The ability of the theory to tell us what to look for in the jumbled remains of the past is one reason the theory is so widely used today by paleontologists.

Biologists have also applied the theory with great success to the question of how birds came into being. Most now agree that flying birds seem to have evolved from one or another small dinosaur. Indeed, the skeleton of the most famous and oldest fossil bird of them all, *Archaeopteryx*, looks very much like that of a little theropod dinosaur. Luckily, the very first fossil specimen of *Archaeopteryx* that was discovered (in 1860) had conspicuous feather impressions, showing that the animal was endowed with essentially modern wings. As a result, there was never any doubt that this specimen was a bird, even though the skeleton in question also had dinosaurian teeth, a long dinosaurian tail, and a simple dinosaurian breast bone incapable of accommodating much in the way of flight muscles.

The connections between dinosaurs and birds were deduced by Thomas Huxley, also known as "Darwin's bulldog" for his tenacious defense of the ideas of his contemporary. Modern evidence in support of the dinosaur-bird hypothesis includes the fact that certain theropod dinosaurs and birds share such distinctive skeletal features as hollow bones, an S-shaped neck, and a wishbone. In fact, paleontologists have now even found a few feathered theropods whose simple "protofeathers," small downy structures, had nothing to do with flight, although they might have provided protection against the cold or perhaps were involved in sexual displays of some sort. In any event, the discovery of a key avian trait, feathers, in what are otherwise clearly nonflying terrestrial dinosaurs pretty much clinches the evolutionary connection between birds and dinosaurs. Thus, the fossil record, imperfect and fragmentary though it may be, provides useful information on the origin of flying birds, eliminating the need to assert incredulously that the ability of these creatures to fly is so special as to require supernatural intervention.

Some opponents of evolutionary theory are apparently aware of the weakness of arguments from personal incredulity and so insist that their criticisms are based not on an inability to imagine how complex traits could evolve naturally. Instead, they claim that they can dismiss Darwinian theory by subjecting it to truly scientific scrutiny, in effect beating evolutionists at their own game. Among these anti-Darwinians are Michael Behe, Philip Johnson, and William Dembski, all academics with university connections. They offer us intelligent design (ID) theory, which they believe better accommodates the biological facts than evolutionary theory. These writers speak of the "irreducible complexity" of living things and the "specified small probabilities" of adaptations arising without external guidance. The ID enthusiasts occasionally salt their commentaries with mention of information theory and references to the philosophy of science. However, the "scientific evidence" that is supposed to be devastating to evolutionary biology really boils down to the fact, not disputed by anyone, that living things are very, very complex, so complex that Behe and company assert that no one can show how standard evolutionary processes could produce these living things. Behe, for example, believes that the elaborate chemical pathways that manufacture a host of useful biochemicals in the cells of everything from bacteria to human beings cannot have been produced one step at a time via natural selection but must have been assembled in a single miraculous event (stage-managed by an unnamed intelligent designer).

Behe and company are, however, really offering us the argument from personal incredulity once again, even though they ornament their position with a certain amount of hifalutin jargon. For example, because Behe cannot imagine how biochemical pathways might change their functions because of fortuitous mutations, he therefore asserts that no such changes were possible. For him and his colleagues, the elaborate adaptations of living things, including the exceedingly complex machinery of life, mean that an intelligent designer must have been involved in their formation.

Biochemists other than Behe have had no such difficulty in accounting for the evolution of complex biochemical pathways and systems in the standard Darwinian fashion. For one thing, geneticists have learned that when cells become sperm or eggs, accidents sometimes occur that cause a chromosome to wind up with two copies of the same gene, rather than the standard single copy. This gene duplication can spread through a species over time via natural selection, provided having two copies instead of one confers even the most modest reproductive advantage on the double gene carrier. Once the doubled copy has become universal, mutations in one of the two copies of the gene will not eliminate the organism's ability to produce the important protein coded for by the other copy of the gene. If the "extra" copy of the gene then enables cells to produce a somewhat different protein that adds a useful reaction to an existing series, individuals so endowed may have more surviving offspring and will therefore leave more copies of the mutant duplication than those with two identical copies of the gene in question. Even minor changes in a gene's base sequence can lead to the production of a protein with a substantially different function, a point that biochemists and geneticists have documented repeatedly. Thus, a mutant duplicated gene with a novel use may initially add a new chemical option for individuals so endowed. But with the passage of time and the incorporation of additional changes in other duplicated genes, what was once an optional gene product could become essential. Currently, the protein products of two very similar genes *must* be produced if you and I are to have a functional hemoglobin in our blood, but other vertebrates get by with just one blood protein of this sort. Almost certainly, when the first duplicated "hemoglobin" gene appeared in a very distant ancestor of humans, the activity of the second gene was beneficial, but not essential, to survival. Over evolutionary time, however, with gradual changes in the two genes, we are now absolutely dependent on this more complex system.

Behe rejects this kind of explanation for complex biochemical pathways out of hand, choosing to focus only on the current highly complex networks of genes and gene products rather than on possible predecessors of these networks. By asserting that these systems are too "irreducibly complex" to have evolved, he creates an unsolvable puzzle in order to call on ID theory. And who might the intelligent designer be? By relabeling the standard creationist, fundamentalist position, Behe and company avoid identifying the God behind ID, which helps their fellow creationists when dickering with school boards and departments of education about the desirability of changing the content of biology courses. If you want to introduce religion into science classes, it is best to claim that your views are scientific rather than religious, given that the courts have ruled that our constitutionally mandated separation of church and state prohibits the teaching of creationism in biology classrooms. Nevertheless, one has to be more than a bit dense not to realize who is supposed to supply the intelligent designing that intelligent design theorists see in the world around them. Johnson, for example, says that he wrote his several books "to redefine what is at issue in the creation-evolution controversy so that Christians, and other believers in God, could find common ground in the most fundamental issue—the reality of God as our true Creator" (*Defeating Darwinism*, p. 92). And Dembski notes with refreshing directness, "Who or what is that intelligence? Within Western culture, it's not a big leap to get to the big G." Although Dembski is circumspect about his religious views, anyone scanning his articles, which are on parade at www.designinference.com, will have little trouble deducing that his enthusiasm for ID theory stems from a religiously based antipathy toward evolutionary theory.

As I write this chapter, the authorities in Ohio are debating whether or not to introduce ID theory into the science classroom. No doubt, twenty or fifty or a hundred years from now, somewhere in the United States the educational powers that be will be grappling with whatever happens to be the latest from dedicated religious lobbyists. The lobbyists will still be insisting that their concerns really have nothing to do with religion, only fairness and evenhandedness in education, the presentation of a full range of ideas in the classroom, or what have you. In contrast, evolutionists then, as now, will be pointing out that ID theory or whatever its latest incarnation is called is really religion in sheep's clothing and should have no place in a science curriculum. Of course, evolutionists are right about the religious nature of ID theory and, if it or any analogue were smuggled into the classroom, at least some creationist science teachers, of which there are some, would be able to proselytize openly under the guise of teaching science. So, strategically, I am sure that evolutionary biologists are right to fight tooth and nail against attempts to insert creationism by any name into the classroom.

But part of me wishes that we could say, Sure, science teachers can discuss ID theory with their students because it offers such a nice example of the difference between a religious theory and a scientific one. Students might get the point once their teachers had a chance to discuss the logic of ID theory, which goes something like this: Because some, but not all, attributes of living things are highly adaptive and extremely complex, the origin and subsequent modification of some, but not all, traits are difficult to explain via Darwinian theory. This means that Darwinian theory is wrong and that therefore, ID theory must be correct.

Now wait just a minute (a science teacher could say). If I say that Bill did not murder Jennifer, does it logically follow that Sam must have done it? Even if one could demonstrate that theory A is wrong, we wouldn't automatically say that therefore theory B is correct, because there may be other explanations not embraced by theory B but available to theory C or theory D. So, for example, I doubt very much that intelligent design advocates would accept the following logic: Some, but not all, attributes of living things are curiously inefficient or seemingly poorly designed, such as the vulnerability of humans to back problems as well as the risk we run of choking to death when food is inadvertently inhaled instead of swallowed. Ugly designs like these are difficult to explain via ID theory. This means that ID theory is wrong, and that therefore, Darwinian theory must be correct.

Because arguments of this sort are devoid of logic, scientists long ago decided that one must actually *test* a theory before declaring victory. Darwinian theory is universally accepted by biologists, not because ID theory has some serious problems, but because Darwinian theory has passed so many tests, thousands upon thousands of them. When fundamentalists say that evolutionists accept Darwinian theory on faith, they are actually partly right, but only because faith in the scientific correctness and utility of the theory is based on truckloads of evidence, as opposed to the blind faith of the creationist in the alternative biblical theory. ID theoreticians have to operate on unquestioned faith because neither they nor anyone else has ever tested the notion that an intelligent supernatural designer was responsible for any aspect of life and the universe. Supernatural phenomena are, by definition, not subject to naturalistic examination, which puts ID theory outside the realm of science.

If high schoolers in Ohio had a chance to learn about ID theory in this light, I suspect no damage and some good would be done by bringing the theory up in the science classroom. I would like to think that the vast majority of high school biology teachers would be able and willing to use the theory to help students understand the distinction between what is and what is not science. But I suppose the fear is that not everyone would play by the rules. No doubt a minority of instructors are so convinced of the correctness of their particular religious perspective that they would turn the biology classroom into Bible camp if given half the chance. So the battle will go on and on to keep the creationist Medusa at bay, despite the time and energy required to chop the head off whatever happens to be the latest attempt by fundamentalists to slip religion into science classrooms.

As we are discussing the relationship between biology and religion, it may be appropriate to examine Darwin's attitudes about this matter. Some historians of science have claimed that Darwin was deeply concerned about the inevitable conflict between his evolutionary view of life and the standard religious position of his time, which was based on a literal reading of the Bible. Adrian Desmond and James Moore highlight this claim in the title of their highly readable biography, *Darwin: The Life of a Tormented Evolutionist*. In their book, Desmond and Moore argue that Darwin's anguish was reflected in (1) the long delay that preceded publication of *On the Origin of Species*; (2) Darwin's reclusiveness, which began as he was developing his theory; and (3) his apparent psychosomatic illnesses, which plagued him for much of his adult life. According to Desmond and Moore, these facts all suggest that Darwin was tormented by an awareness of the social implications of his scientific work.

And, yes, Darwin knew that both the religious establishment and his religious colleagues were unlikely to receive his theory with much pleasure. After all, during his undergraduate days at Cambridge, he was technically in training to become a clergyman. Moreover, his wife was a devout person who told him of her discomfort with certain of his scientific views. But for someone who was supposedly eager to avoid the social pressure and family unhappiness that would inevitably come from presentation of the theory of natural selection, Darwin lays down some remarkably direct challenges to creationism in *Origin* and in his other books as well. For example, in the last chapter of *Origin* he repeatedly draws the reader's attention to particular facts that are incompatible with the notion that "God did it." To pick one small example, Darwin notes that bats occur on many oceanic islands that are otherwise unoccupied by terrestrial mammals. This finding is hardly surprising if oceanic islands could only be colonized from neighboring continents by mammals capable of flying across large stretches of water; the only such mammals are bats. In contrast, if God had the infinite power to populate the various corners of Earth with species of his choosing, then the bats-only policy on oceanic islands makes little sense. Surely He would have wanted all large islands to have a terrestrial mammal or two, but they are not there. And Darwin does not make it difficult for his reader to catch his drift here and on allied issues when he makes statements such as, "Such facts as the presence of peculiar species of bats, and the absence of all other mammals, on oceanic islands, are utterly inexplicable on the theory of independent acts of creation" (pp. 477–478).

Darwin's disdain for the creationist argument from personal incredulity appears in the following quote, also from *Origin's* last chapter: "When we no longer look at an organic being as a savage looks at a ship, as something wholly beyond his comprehension; when we regard every production of nature as one with a long history . . . how far more interesting—I speak from experience—does the study of natural history become!" (pp. 485–486). Here he manages to put in a plug for evolutionary theory while equating the opposition with acceptance of an intellectual position that might appeal to an uncivilized savage, a hard-hitting comparison for his British readers, convinced as they were of their superiority over the aboriginal peoples who populated much of the extensive British Empire at the time.

In a later book, *The Descent of Man and Selection in Relation to Sex*, Darwin tackles the human question head on when he writes that evolutionary theory enables us to "understand how it has come to pass that man and all other vertebrate animals have been constructed on the same general model, why they pass through the same early stages of development, and why they retain certain rudiments in common. Consequently we ought frankly to admit their community of descent; to take any other view, is to admit that our own structure, and that of all the animals around us, is a mere snare laid to entrap our judgment. . . . It is only our natural prejudice, and that arrogance which made our forefathers declare that they were descended from demi-gods, which leads us to demur to this conclusion" (p. 32). I fail to see much hesitation or reserve in this paragraph, which offers a clear rebuke to that subset of creationists who claim that God endowed humans with certain animalian features solely in order to trick (entrap) us into thinking that there was an evolutionary connection between humans and other living things when there was none.

Likewise, when Darwin begins his book on orchids, he tells his reader that he will attempt to "show that the study of organic beings may be as interesting to an observer who is fully convinced that the structure of each is due to secondary laws, as to one who views every trifling detail of structure as the result of the direct interposition of the Creator" (*The Various Contrivances by which Orchids are Fertilised by Insects*, p. 2). By "secondary laws," Darwin refers to the natural laws of physics, chemistry, and biology, which operate free from supernatural intervention.

As discussed already, Darwin then proceeds to examine one species of orchid after another, working through the adaptive puzzles provided by the elaborate and often bizarre flowers of these plants, showing over and over again that almost every structural element of a given species has a useful role to play in getting that orchid pollinated by an insect. Having made this point almost to exhaustion, Darwin then turns to issues of origin, rather than adaptation, by bringing to our attention a few elements of certain orchids that currently have no function. He argues that these features are vestigial structures, remnants of traits more fully developed in an ancestral species from which the current orchid species in question has evolved. For example, in *Catasetum* (a genus in which there are separate male and female flowers), the female flowers have rudimentary, nonfunctional (male) pollen masses, clearly a vestigial trait inherited from an ancestor in which both functional pollen-producing and pollen-receiving structures were present in the same flower. That the ancestor of *Catasetum* produced hermaphroditic flowers is entirely plausible given that this kind of sexuality characterizes the overwhelming majority of all modern orchids. Having made this point, Darwin writes, "At a period not far distant, naturalists will hear with surprise, perhaps with derision, that grave and learned men formerly maintained that such useless organs were not remnants retained by inheritance, but were specially created and arranged in their proper places like dishes on a table (this is the simile of a distinguished botanist) by an Omnipotent hand 'to complete the scheme of nature'" (*The Various Contrivances by which Orchids are Fertilised by Insects*, p. 203).

Darwin's comments on *Catasetum* are those of someone who feels entirely comfortable tackling fundamentalist views on the diversity and adaptedness of life. In fact, far from backing away from a confrontation, he almost seems to be inviting a dustup. Although Darwin did rather grudgingly accept the possibility that an Omnipotent Creator may have gotten the ball rolling by forming the initial living thing, a single-celled starter organism, that was that. As far as Darwin was concerned, all the species currently on Earth are the modified products of evolution by natural processes. He did not muddy this conclusion nor seek some sort of middle ground. Nor did he offer the fundamentalist an easy way out. All of which leads me to question whether Darwin's psyche really was as fragile as it is sometimes portrayed. Of course Darwin realized that his ideas were going to make a lot of people unhappy, but in the last analysis he did not flinch from presenting his sacrilegious ideas in a way that everyone could understand.

As an orchid enthusiast and evolutionist, I like the fact that Darwin used orchid diversity (in the sixth edition of *On the Origin of Species*) as a lead-in to his rhetorical question, "Why, on the theory of creation, should there be so much variety and so little real novelty?" (p. 270). The answer was clear to Darwin: despite great differences among orchid flowers, all members of this group nevertheless share certain distinctive attributes. Moreover, many living species of orchids exhibit transitional stages of flowers present in more complex form in other species, evidence that shows how a more complicated feature could arise step by step from a less complicated ancestral type. The abundance of variations on a central theme and the extreme scarcity of species with apparently unique attributes simply doesn't make sense if each species was created independently of all others by an all-powerful God limited only by His or Her imagination. In contrast, the fact that variety far exceeds novelty is entirely consonant with the evolutionary theory of descent with modification, in which an ancestral species endows its many descendant species with key structures that then become modified to varying degrees over time. Darwin knew that a natural, not supernatural, history of a particular sort could explain the origin of species, and he wasn't about to conceal what he knew.

Darwin's studies of orchids are relevant not just to the creationist-evolutionist debate but also for arguments that occur entirely within the circle of committed evolutionists who have no patience with creationist machinations. One such truly scientific argument has to do with questions of origin and modification, with Stephen Jay Gould, that chastiser of adaptationists, also taking up the cudgels against what he felt was the defective thinking of those who were interested in the history behind modern attributes of living things. To this end, he joined forces with Elisabeth Vrba in claiming that many biologists failed to grasp the difference between traits that were shaped by natural selection for their current function versus those that had been "co-opted" for a new use. In their 1982 paper, which had been cited in academic research papers more than seven hundred times as of 2004, Gould and Vrba state that "adaptation" should be strictly reserved for those traits that have retained their original adaptive function. They claim that everything else deserves a new title: *exaptation*. So, for example, because the wing feathers of birds that are so essential for avian flight almost certainly were preceded by downy insulating feathers, flight feathers would be exaptations in the Gould and Vrba lexicon because the current function of these feathers differs from their original adaptive utility.

The effect of Gould and Vrba's semantic suggestion, if accepted, would be to reduce the number of "adaptations" in this world. To make this point clear, they assert that adaptations have (evolved) functions, whereas exaptations only have (side) effects. In other words, persons interested in the evolved purpose of a trait had better first determine whether they are looking at something whose adaptive role had not changed over evolutionary time. Otherwise, they might wind up assigning adaptive significance to something that really needed to be explained as a nonadaptive by-product of history. To provide support for their argument, Gould and Vrba invoke Darwin by citing his comments on the moderately arcane subject of mammalian skull sutures. Here is what the father of evolutionary theory had to say on this matter: "The sutures in the skulls of young mammals have been advanced as a beautiful adaptation for aiding parturition, and no doubt they facilitate, or may be indispensable for this act; but as sutures occur in the skulls of young birds and reptiles, which have only to escape from a broken egg, we may infer that this structure has arisen from the laws of growth, and has been taken advantage of in the parturition of the higher animals" (first edition of *On the Origin of Species*, p. 197).

Gould and Vrba claim, on the basis of this paragraph, that Darwin believed that an adaptation was something that evolved for a particular purpose. Therefore, skull sutures in young mammals were out because these features, although presumably useful during the process of birth, had originated for some other reason, perhaps as side effects of the "laws of growth." In other words, mammalian skull sutures came about because of the old-fashioned way that some developmental systems work, systems that mammals have retained from their reptilian ancestors. If true, Gould and Vrba would have us agree that we need not investigate the adaptive value or evolved function of these developmental by-products of history.

There is no doubt that Darwin understood the distinction that Gould and Vrba wish to make between adaptation and exaptation. For example, in his orchid book Darwin writes, "When this or that part has been spoken of as adapted for some special purpose, it must not be supposed that it was originally always formed for this sole purpose" (*The Various Contrivances by which Orchids are Fertilised by Insects*, p. 282). A case that Darwin uses to illustrate his argument involves the elastic stalk (or pedicel) that attaches to the pollinia in *Catasetum* orchids. The pedicel acts rather like a catapult to fling a pollen mass at a pollinator at high speed after the insect has inadvertently touched the antennae of the flower, as described earlier. According to Darwin, this quite astonishing catapult was derived from a much less dramatically elastic stalk present in the past, a device that had the capacity to bend quickly, but not hurl itself through space, once the stalk had been touched by a pollinator. In a predecessor to *Catasetum*, the rapid movement of this pedicel probably pulled the pollen masses from their containers in the anther so that they could be picked up by an insect pollinator. This useful function can be performed by a rod capable of bending, rather than catapulting. This kind of device need not generate the force needed to throw the pollen masses through the air at 323 centimeters per second.

Thus, as far as Darwin was concerned, the *Catasetum* pedicel was not originally formed for its current specialized function, which it has acquired through modifications of a different kind of pedicel that had another purpose at an earlier time. And this is the common pattern, according to Darwin, who goes on to say, "The regular course of events seems to be, that a part which originally served for one purpose, becomes adapted by slow changes for widely different purposes" (*The Various Contrivances by which Orchids are Fertilised by Insects*, p. 282).

Darwin applied the same perspective in his studies of complex plant behavior. So, after having established how certain climbing plants were able to twine their way upward around a supporting stem, he predicted that this sophisticated response was preceded by a simpler trait with another function. When he went to look for it in ordinary, nonclimbing plants, he discovered that the growing stems of many of these plants engaged in slow and subtle actions, during which a stem tip "bends successively to all points of the compass" (*The Power of Movement in Plants*, p. 3), perhaps enabling the plant to adjust the position of its terminal leaves or flowers in relation to sunlight. To investigate these micromovements, he developed a most elaborate methodology, which enabled him to record plant tip movements of as little as 1/500 of an inch. Darwin gave the phenomenon the name *circumnutation* (a nice bit of jargon welcomed by botanists and still in use today), and he wrote a book about the huge battery of experiments he conducted on plant movements with the assistance of his son George. I confess that reading the book is almost as riveting as watching a runner bean grow, although Darwin's account does give you a feeling for the meticulous nature of his work and the productivity of his dedication to the proposition that complex adaptive traits with certain current functions must have originated as simpler traits with other functions.

Perhaps because of Darwin's writings, an awareness of the distinction between the origin of a trait and its current function has been standard for evolutionary biologists for many years. Even when I was a student, a depressingly long time ago, I learned that some structural features originate with one purpose but then acquire another as time passes. One of the classic cases, which I had to memorize in the 1960s, involves the three little ear bones (hammer, anvil, and stirrup) that endow us and many other vertebrates with the capacity to hear. Two of these bones (the hammer and anvil) have been derived from two small bones that once were located in the lower jaw of reptiles. The original role these bones performed in jaw structure and movement obviously had little to do with the sophisticated acoustical services that vertebrate ear bones provide today.

Many other examples exist, as we have already demonstrated with such cases as the likelihood that the false sex pheromone of some European orchids evolved from the waxy waterproofing of the orchid flower or the flycatching glue of the sundew having evolved from an herbivore repellent. The more general point, emphasized by Kern Reeve and Paul Sherman, two of Cornell University's evolutionary biologists, is that if one goes far enough back in time, almost all of today's adaptations would be shown to have originated with a different form and function. In other words, Gould and Vrba's exaptation would have to be applied to almost all of today's adaptive traits. This means that exaptation and adaptation are essentially synonyms, and if true, then the new word is unnecessary. I join Reeve and Sherman in thinking that we biologists have more than our fair share of jargon already and so are in no need of any supplements.

My belief that Darwin would surely agree with Reeve, Sherman, and me rests in part on the following quote, in which Darwin almost seems to be anticipating Gould and Vrba's argument: "Although an organ may not have been originally formed for some special purpose, if it now serves for this end, we are justified in saying that it is specially adapted for it" (*The Various Contrivances by which Orchids are Fertilised by Insects*, p. 283). In other words, that organ or feature can be called an adaptation and be done with it.

Darwin adds, "On the same principle, if a man were to make a machine for some special purpose, but were to use old wheels, springs, and pulleys, only slightly altered, the whole machine, with all its parts, might be said to be specially contrived for its present purpose. Thus throughout nature almost every part of each living being has probably served, in a slightly modified condition, for diverse purposes" (*The Various Contrivances by which Orchids are Fertilised by Insects*, pp. 283–284).

The point, laid out so clearly by Darwin, is that living things are the jury-rigged products of past selection, which operated on the heritable bits and pieces that were available in ancestral populations. What we see today in an orchid or a bird or a human being are a host of adaptive attributes that spread in the recent past or have been maintained over time because they made a greater contribution toward the reproductive success of individuals than other alternatives that happened to arise by chance within a species from time to time. If today's traits are being retained because of the reproductive advantage they currently offer over other variants, then these features of a species can be called adaptations, whether or not the function they happen to serve now is the same as that achieved by their predecessors earlier in the history of the species.

Evolutionary history is in large measure a history of adaptations. Descent with modification is the means by which an adaptive trait, once having originated through a chance alteration of an existing attribute, spreads through an entire species. Should a slight variant of the new norm arise with superior effects on individual reproductive success, then it, too, can spread through the species over time. As one better (almost certainly never perfect) trait replaces another through the action of natural selection, the history of the species is written in its transient adaptations. Let us thank the process for having produced the warty hammer orchid and thousands of other orchidaceous delights.

As I reexamine my well-worn field guide to the orchids of southwestern Australia, I read that the Esperance flying duck orchid is the least known of the flying ducks, eight species of which grace the pages of Hoffman and Brown's book. The orchid had not even been formally named at the time of publication of this book, which reports that the species flowers in November (the end of the Australian spring) and occurs only in a couple of places well east of Esperance, a small port city a very long day's drive to the east of Perth. Based on the orchid guide's information, I know that I do not have a realistic chance of ever seeing this orchid, but I would like to try, just the same. Hoffman and Brown give me a splinter of hope to cling to by noting that two well-known amateur orchid hunters, Gloria and the late Bill Jackson, found a few small colonies of the orchid off the Balladonia Track in 1999 in the same general area where Andrew Brown had found the species a couple of years earlier. This road heads north through the inland portion of Cape Arid National Park, that most lonesome of wilderness parks. Well, if the Jacksons have been able to locate the Esperance flying duck along the Balladonia Track, perhaps I can, too.

Long before I decided that I wanted to hunt for the Esperance flying duck, I was intrigued by the Balladonia Track, which appears on maps of Western Australia as a narrow dotted line running off into a kind of no-man's land of empty space unmarred by town, hamlet, river, or other road. After 50 kilometers, the track reaches an isolated peak, Mt. Ragged, where the road hooks up with another narrow dotted line out in the middle of Nowhere with a capital N. The track then turns due north and

5

Orchids, Species, and Names

runs for 150 kilometers or so before it eventually reaches the main east-west road coming from the Nullarbor Plain. The road across the Nullarbor is called the Eyre Highway, and it is kept in excellent shape, but even on this primary roadway traffic is very light, as befits one of the loneliest, least populated regions of Australia.

Thus, the Balladonia Track offers an alluring ticket of entry into a relatively rarely visited part of Australia. But, and this is a big but, the maps I studied all indicated by their narrow, dotted lines that the Track was for four-wheel-drive vehicles only. My wife and I travel in Australia in a two-wheel-drive campervan, a wonderful machine in many ways but utterly hopeless in sand or mud, as bitter experience has taught me. I knew that if I were to head off down the Balladonia Track in the campervan, we would be up to our axles in sand or mud before we had stopped singing the first verse of "Waltzing Matilda." Therefore, I had more or less resigned myself to a life without experiencing either the Track or the Esperance flying duck—until Leigh Simmons, a biologist at the University of Western Australia, suggested that he, his family, and his Subaru rendezvous with us at Cape Arid National Park. Leigh bragged that his Forester was a superbly competent four-wheel-drive sedan, one that could in theory transport all five of us down any four-wheel-drive track of our choosing. I leapt at his offer. In due course, Leigh, Carol, and Freddy made the trip to Cape Arid, where they joined us in a beautiful spacious campsite overlooking Yokinup Bay. From our perch we used our binoculars to keep tabs on a female southern right whale and her calf wallowing in the gentle swell just outside the breakers that swept the beach clean day and night.

The morning after our companions had arrived at Cape Arid, all five of us piled into the Subaru for our four-wheel excursion. Our short zig-zag journey along good dirt roads out of the park went quickly except for a brief delay caused by an encounter with a mob of sheep being driven down the road by two men who were moving their flock from one field to another. After inching through the massed sheep, we soon came to the turnoff to the Balladonia Track. There a weathered sign, slightly askew, announced that the way ahead was, as we already knew, strictly for four-wheel-drive vehicles and then only when the road was dry. The first few kilometers of track had been upgraded recently, and so we had a brief reprieve before reaching the start of the real thing. There another, less-battered sign greeted us with another warning: Deep sand, muddy stretches, and rocky outcrops ahead. As I surveyed the rutted white sand immediately in front of us, I could see that the sign makers were not just making it up, and I wondered if what we were about to do was wise. The thought of four-wheeling into the wilderness had been enticing, but the view of the actual deeply rutted, one-lane track ahead provided something of a reality check. I could feel my mild sense of excitement fading, and in its place, a moderate dose of anxiety took root. But Leigh did not hesitate. The Forester plunged into the deep furrows of sand and struggled doggedly forward, slowing on occasion but never coming to a complete halt.

Soon I stopped imagining how unpleasant it would be to dig ourselves out if we did become sand-trapped. Instead, I turned my attention to the low heath through which we were slowly but steadily advancing. With the opportunity to reflect on the main purpose of our journey, my thoughts returned to flying duck orchids, which do well in sandy habitats. So I suggested that if we could find a stretch of the road with reasonably hard-packed sand, we should stop to take a look around. A suitably firm spot soon presented itself, and out we went to do some serious orchid hunting along an overgrown side track that angled off the main road.

The day had begun under heavy overcast. Now a light drizzle started to fall. I tried not to think about the warning that the track would become impassable when wet in order to concentrate on the task at hand. As a realist prone to pessimism, I always believed that our chances of finding the orchid here or anywhere else along the Balladonia Track were as close to nil as they could be without hitting absolute zero. Bill and Gloria Jackson were not only vastly more experienced orchid hunters than we were, but they could have come across the orchid anywhere between the start of the track and its terminus near Mt. Ragged, 50 kilometers away. Moreover, inasmuch as it was only the first of November, very early in the supposed flowering period of the species, the orchid might very well not be in bloom even if we were fortunate enough to find the object of our desire.

After about twenty minutes of careful scanning in the little bare patches beneath and by the scruffy shrubs growing up in the old track, I gave up and turned back toward our trusty Subaru, having concluded that we were not going to have the thrill of finding the Esperance flying duck orchid here. My glasses had become misted with drizzle. Leigh then announced, "I have one." His voice combined surprise and pleasure, and rightly so, as I was able to see for myself after rushing over to stare at a skinny three-inch high stalk with a deep red bud on top. Alas, the orchid was indeed only in bud, not in flower, and therefore not suited either for identification or for photography. But with clear evidence that this was flying duck habitat, the four adults in our company turned to the search with renewed vigor and considerable excitement. Leigh was first to find a specimen in full flower, and between us we turned up a dozen or so plants either in bud or with an opened flower just asking to be photographed. I obliged.

Because the orchid field guide stated that only the Esperance flying duck orchid occurred in the Cape Arid region, all of us assumed that we had succeeded against all odds in finding this species on our very first Balladonian attempt. We congratulated ourselves, and Leigh in particular, on our collective good fortune. Then, after taking one last photograph of the orchid, it was back in the Subaru for the rest of the trip, which featured some large and thoroughly intimidating lakes of brown water that covered the track from side to side for several hundred meters. However, the Subaru swam through these obstacles without a problem, just as it conquered, ever so slowly, the horribly rocky stretches that we came upon later. Although we stopped to search for more flying ducks here and there, we, or I should say Carol, found only a single additional specimen, also in bud. The trip out and back, although barely more than 100 kilometers, took us the full day, what with our pauses to hunt for orchids or to somberly survey the track's muddy segments or crawl over the small boulders and stone ridges that made rapid travel impossible near Mt. Ragged.

The next day was sunny and so better suited for camera work. Therefore, I talked Leigh into revisiting the orchid site, which required renegotiating only the first few miles of white sand, once we reached the four-wheel-drive section of the Balladonia Track. There we found still more flying duck orchids in flower, and I set about adding to my slide collection. As I was crouching down to get close enough to photograph a particularly nice specimen, I realized that its labellum looked different from that of its immediate neighbor, which was also in bloom. This neighboring plant, and our earlier finds, all had small lip petals, with an irregular smattering of dark calli at the very tip. In contrast, the distinctive individual right in front of me had a somewhat larger labellum with a dark mass of calli neatly covering the outer third of bill. I was puzzled. Leigh was puzzled, too. Later on, I found another orchid with the same substantial band of calli growing beside others with a few terminal calli.

Back to the orchid book we went. There we learned that the Esperance flying duck closely resembled the little flying duck, *Paracaleana linearifolia*, another species that occurs in scattered populations much

5.1
Two flying duck orchids of Cape Arid National Park. (top) A plant with a very few dark calli concentrated on the very tip of the orchid's lip petal. (bottom) A specimen with a considerably larger, much more sharply delineated band of calli on the duck's "bill."

closer to Perth, far to the west of Cape Arid. The little flying duck orchid, which also flowers in November, not only is smaller than most of its relatives, but it also possesses a clean-cut dark band of calli that cover the distal third of the labellum. In other words, the two dark-banded specimens that I had found on our second visit to the Balladonia Track were almost certainly the "real" Esperance flying duck orchid. But what about the ones with many fewer, more terminal calli? Were these just unusually small specimens of this same species with a correspondingly reduced batch of calli? Or were they another species altogether?

To resolve this mystery, I arranged a meeting with Andrew Brown, who looked at my photographs and said, "So, you found that one." He then told me that we had stumbled upon a flying duck orchid that had been discovered so recently that it neither had a name nor a photograph in the revised edition of the orchid field guide. He was confident that two species were involved because he knew that the flying ducks along the track could be cleanly divided into two groups, one with a large, sharp band of calli on the labellum tip and the other with a miserly, irregular set of black bumps at the very end of the labellum.

These kinds of differences in appearance, modest though they may be, are the basis for identifying the various species of flying ducks in Western Australia. Not so many years ago, only a single species, *Paracaleana nigrita*, was recognized from this entire region. But then, as orchidologists, especially Andrew Brown and Steve Hopper, began to study the plants more closely, they found that their botanical predecessors had overlooked subtle but real differences among populations with respect to the shape of the duck-billed labellum and its calli. Thus, the flying ducks that currently go by the scientific label *Paracaleana nigrita* all have a narrow humped labellum unlike that of any other currently recognized species (see Figure 2.11, top). In addition, the little black-purple calli that decorate the humped labellum of this species cover about half the upper surface of this petal, whereas in other species, the distribution of these glands differs. For example, the narrow, flat labellum of the smooth-billed flying duck, *Paracaleana terminalis*, has only a narrow black band of glandular tissue on the tip of the bill. In *Paracaleana triens*, the broad-billed flying duck, the calli cover the distal third or so of the bill, not just its tip, and the bill itself is much wider at its base than is true for either *P. nigrita* or *P. terminalis*.

Then there is Hort's flying duck, another species yet to be formally named, whose existence was recognized only in 1997 thanks to Fred and Jean Hort, another Australian couple who have been bitten by the orchid bug. The Horts realized that some flying ducks found in the hills to the east of Perth had a flat, not humped, labellum that tapered into a thin rectangular point, not the rounded bill tip of all other flying ducks. Because of the late date of this discovery, a photograph of Hort's flying duck did not appear in Hoffman and Brown's field guide until a revised edition was published in 1998. When I saw this version in 1999, I decided (of course) that here, too, was a species that I just had to see in person. To this end, Sue and I spent an October afternoon driving around searching for appropriate habitat, which was said to be sandy patches in the open eucalyptus woodland located near York, a small town inland from the big city, Perth. Unfortunately, I had managed to misread the guide, and so searched energetically in an area to the *east*, not the west, of York. By the time I realized we were barking up the wrong tree, it was too late to remedy my mistake during that year's visit to Western Australia, and I had to carry on without having seen Hort's flying duck orchid in 1999.

We returned to Australia in 2001, and this time, I had my geography in order with respect to where Hort's flying duck could be found. Therefore, Sue and I invested another day driving through the woodland between Perth and York, a forested region that has not been cleared, in part because it serves as a water catchment area for the big city. But even though we knew roughly where the orchid had to be, finding the darn thing proved to be a challenge. As far as I could tell, Hort's flying duck should have been in all the forest patches where we stopped to search. The habitat looked ideal, with its scattered big trees, shrubby understory, and sandy patches, one seemingly perfect spot after another. And yet, all these sites proved to be devoid of the orchid in question, although we did see donkey orchids here, purple enamel orchids there, and lemon-scented sun orchids almost everywhere, discoveries that helped keep our attention focused and our spirits up.

After any number of stops, I drove down what was surely an old logging track and parked by a large sandy clearing dotted with scattered shrubs and sedges and some lonely looking eucalypts. Soon I found a hammer orchid, a good sign because flying ducks and hammer orchids often grow together, a point that I confirmed soon thereafter when I spotted a flying duck in a sandy patch by a blackened log, incinerated no doubt during one of the controlled burns so favored by Australian forest managers. I very much wanted this specimen to be Hort's flying duck, but sadly for me, I could not flatten out the obvious hump in its labellum, no matter what angle I chose for my inspection. I had found a common flying duck orchid, *Paracaleana nigrita*. Only a little time passed before I found some more specimens of this same species, at which point I encouraged Sue, who had grown weary of orchid searching, to come look at the flying ducks and hammer orchids I had found. She obliged with even more controlled enthusiasm than usual, and then had a look around herself. Before long, she called me over to a little stand of five flying duck orchids.

Eureka! There they were: Hort's flying duck orchid with the flat, not humped, labellum with a rectangular, not rounded, point to the bill. I was thrilled and properly grateful for my spouse's assistance, and I told her so. Yet had I not been looking specifically for Hort's flying duck with knowledge of the key features that distinguish it from the common flying duck, I am confident that I would have briefly admired the cluster of plants that Sue had found, approving of the find but then moving on. I might have commented that the orchids looked a little odd, but almost certainly I would not have realized that I had been looking at a different species.

All of which raises some fundamental evolutionary questions that are on a par with questions about adaptations and their origins: Just what is a species? How do they come into being? And why were the Horts sure that they were looking at a different species when they first stumbled across Hort's flying duck? With respect to the definition of a species, you might think that biologists would have managed to get this matter straight by now. After all, the scientific procedures for naming species were put in place in the mid-1700s by Karl Linnaeus. This Swedish biologist invented the most useful system of giving distinctive two-word names (in Latin) to each species, as in *Homo sapiens*, the scientific name of our species, and *Paracaleana nigrita*, the scientific name of the common flying duck orchid. All scientific names begin with the generic label (e.g., *Homo*) and end with the specific title (e.g., *sapiens*). Although, as we have seen, more than one species can be placed within the same genus, every species within a given genus has its very own specific name. As a result of having a single Latin name for each species rather than a host of common names in a variety of languages for one and the same species, a great deal of confusion can be avoided.

5.2

Hort's flying duck orchid, another species in the genus *Paracaleana*, which can be distinguished from the common flying duck by its flat labellum with an almost rectangular tip.

Admittedly, Linnaeus developed his naming system at a time when many others had their own lists of names. According to his biographer Lisbet Koerner, his scheme caught on primarily because it was relatively simple and accessible to students, amateurs, and academics alike. Whatever the reason, Linnaeus has received much credit for his procedures, and deservedly so, because thanks to him, people the world around can communicate clearly about species X, Y, and Z.

Linnaeus published the first installment of his classification system when he was just twenty-eight, not long after he had completed his medical training. Because he was a doctor, he was doubtless aware of the benefits of being able to talk with others about the precise identity of plants of medicinal value, of which there are many. In his early years as a practicing doctor, Linnaeus surely employed herbal cures for the patients who came to him with syphilis. At the time, soapwort (*Saponaria officinalis*) was considered especially effective against the disease, as were many other plants, including woody nightshade (*Solanum dulcamara*), saffron crocus (*Crocus sativa*), and sand sedge (*Carex arenaria*), all species for which Linnaeus himself supplied the Latin names that are still in use today.

Given that Linnaeus viewed his taxonomic system primarily as an economic tool, I imagine that he was aware of the medicinal virtues of certain orchids, which were once prescribed to cure impotence or to induce sexual restraint, depending on the doctor and the patient. Linnaeus wrote several technical papers on orchids, describing eight genera and at least fifty-nine species in the process. Ever since

Linnaeus, orchidologists have been using the Linnaean system to good effect. The common flying duck *Paracaleana nigrita* came to the attention of botanists in the first half of the nineteenth century, when the Australian botanical entrepreneur James Drummond included specimens of the orchid in shipments sent to Europe for sale. Drummond described the flying duck in a letter published in 1840 in an English botanical journal with the assistance of John Lindley, a botanist long associated with the British Horticultural Society who was also responsible for the generic name of the hammer orchids (*Drakaea*).

Subsequently, Steve Hopper and Andrew Brown reexamined some populations of flying ducks that were then considered to be *Paracaleana nigrita*, and, as mentioned, they recognized species that had been overlooked previously, such as the smooth-billed flying duck orchid. The common English name "smooth-billed flying duck" has no scientific standing but is available for those of us who wish to avoid memorizing the Latin label. The name that really counts, however, is *Paracaleana terminalis*, the name given to this species by Hopper and Brown because of the terminal cluster of calli ornamenting the extreme end of the labellum. This scientific name is the one that botanists now can use when they wish to refer to the species in a way that makes it unmistakably clear just what plant they are talking about.

Linnaeus and most other taxonomists subsequent to him, including Hopper and Brown, have given unique scientific names to orchid populations that differ consistently in appearance from one another. The critical features for the flying duck orchids, as we have seen, include the shape of the labellum and the distribution of the calli on this flower petal. As far as I know, everyone agrees that *P. nigrita* and *P. terminalis* are separate species, thanks to their unique characteristics that make it possible to place individuals into one or the other population without ambiguity. We all know, however, that a single species can be composed of subpopulations that differ in appearance to a much greater extent than *Paracaleana nigrita* and *Paracaleana terminalis*. Take the human species, for example. If a Martian taxonomist were to examine three samples of human beings, one group of Australian aborigines, another of an indigenous Amazonian tribe, and a third taken from downtown Tokyo, he or she or it might conclude that the three groups belong to three different species. But Australian aborigines, Amazonian Jivaro, and the inhabitants of Tokyo are all currently classified by Earth-bound taxonomists as belonging to the same species, *Homo sapiens*.

Why? In part because given a large enough sample of people, one can find individuals who are intermediate in appearance to others, which helps tie the various subpopulations together in terms of their external features. More important, ample evidence exists that humans the world around can copulate with one another, an activity that is sometimes followed by the production of viable offspring. This simple fact demonstrates the genetic compatibility of the different races, or whatever you choose to call the groups of humans that live in different parts of the globe. The shared genetics of Australian aborigines, Jivaros, and Japanese are more than sufficient to enable individuals from the various groups to reproduce together despite their physical differences. Over human evolutionary history, a combination of occasional matings at the edges of different populations and the absence of large genetic changes even in well-separated groups has maintained the genetic cohesiveness of our widely dispersed species. In other words, a species can be thought of as a shared gene pool that has evolved in isolation from other collections of genes, thanks to the reproductive mechanisms that determine who can exchange genes with whom.

The special genetic features that enable a man and a woman to have children, no matter that their ancestors were very different, not only pull our diverse populations together, but they also prevent us from reproducing successfully with members of other species, which are equally disinclined to treat us as potential mates because of their different genetic heritage. Chimpanzees and gorillas are, as noted already, remarkably similar genetically to human beings despite the fact that they do not look much like us. Certain genetic differences between us and them must generate the obvious external differences that distinguish us from them. Other unique genetic features are responsible for assorted internal differences in the skeletons, digestive systems, and brains of the different species. A combination of all these factors creates the barriers that keep us apart reproductively, with the result that the gene pools of humans, chimps, and gorillas have not merged over evolutionary time.

Therefore, when it comes to knowing whether a group of populations deserves to be split into a set of separate species, as opposed to being lumped together under the banner of one species' name, many biologists like to have direct or at least indirect evidence about the degree of reproductive isolation between the populations under study. (Please note that not all biologists believe that "species" is synonymous with "reproductively isolated population." Indeed, a good many botanists put more weight on the ability to discriminate species by virtue of shared appearance, biochemistry, or some other attribute that is more conveniently measured than reproductive isolation.)

For those who do focus on reproductive separateness, however, failure to document reproductive barriers between two populations is grounds to doubt their separate identities. Thus, most biologists came to discount the claims of the prolific mammalogist C. Hart Merriam, who gave eighty-six specific names to eighty-six populations of grizzly bears living in Mexico, the United States, and Canada. In his orgy of species naming, Merriam argued (for example) that four species of grizzlies made Colorado home, and five species could be found in Yellowstone, all on the basis of no more than modest differences in size and coloration. Subsequently, other biologists overruled Merriam by pointing out that the different populations graded imperceptibly into one another, strongly suggesting that grizzly bears throughout North America were able and willing to interbreed. If so, they could not be separate species, if one used the criterion of reproductive isolation. The likelihood that North American grizzlies constitute one wide-ranging, interbreeding population has received additional support from modern genetic studies that show, for example, that the very large brown bears of coastal Alaska are not genetically distinguishable from the considerably smaller grizzlies of the northern interior. Instead of eighty-six species of grizzlies, most researchers believe that just one species occurs in North America.

The same sort of thing has happened more recently in the world of Australian orchids. One such orchid occurs all the way from northern Queensland south to Victoria along the coast. Although different populations share obvious similarities, the flowers vary markedly in size, shape, and color (from white to gold) over this wide distribution, generating much discussion among the experts on how to classify the various forms, which occupy different parts of the geographic range. Some botanists were in favor of considering the different types to be variants within a single species, but others were on record favoring the division of the group into five species. Then, in 2002, Jacinta Burke and Peter Adams published the results of an exhaustive analysis of hundreds of specimens from all parts of the range, involving thousands of measurements of floral and vegetative characters. The resulting numbers were subjected to assorted tests to determine if the members of some populations could be statistically separated from one another. They could not, leading Burke and Adams to conclude that these orchids deserved to be considered members of a single species, *Dendrobium speciosum*, which had undergone a degree of geographic differentiation but not so much as to warrant recognizing a cluster of truly different species. The fact that intermediate forms could be found between all the populations suggested that genes were being exchanged among them, one reason for considering them all part of one single interbreeding unit. This may or may not be the last word on the orchid, but at least now we have a carefully considered rationale for lumping the many different populations into a single species.

On the other hand, lumpers sometimes have probably gone overboard. The African elephants were once divided into more than a dozen species, but then were coalesced into just one species, *Loxodonta africana*. Subsequently, however, a team of researchers collected tissue from wild elephants living throughout Africa (by shooting biopsy darts into their subjects). After no doubt cautiously retrieving the darts and analyzing the DNA extracted from the flesh within, they concluded that there were two genetically distinctive clusters of elephants, one associated with African forests and the other with African savannahs. Each group of elephants had its own forms of the genes in question, with almost no evidence of gene flow between the two populations, suggesting considerable reproductive isolation between them. This fact, coupled with certain differences in their tusks, ears, and skulls, convinced Alfred Roca and his colleagues that they were dealing with two species, not one. Thus was born *Loxodonta cyclotis*. More recently still, additional DNA fingerprinting tests (applied to DNA extracted from elephant dung collected in different parts of the elephants' range) have convinced other mammalogists that a third species of elephant needs to be formally recognized.

Elephants are not the only animals in which species multiplication via taxonomic splitting has occurred. At one time, the medium-size mottled brown and green frogs found throughout much of North America were lumped together in what was considered to be a widespread and variable species, the leopard frog *Rana pipiens*. But then John Frost and Jim Platz took a closer listen to the calls produced by males of supposed *Rana pipiens* only to find that the come-hither songs of males differed greatly. Males in one population produced a staccato-snore, in another, a trilling call, and so on. These differences in courtship signals meant that males of population X were most unlikely to attract and mate with females of population Y, thereby blocking the exchange of genes between the two populations. If indeed the populations were reproductively discrete, then they should also possess distinctive genetic information; genetic tests revealed, as predicted, that each population had unique forms of any number of genes. The genetic and behavioral differences meant that the frogs in population X almost certainly did not interbreed with the frogs in population Y. As a result, herpetologists now recognize seven different species of leopard frogs, including *Rana pipiens*, *Rana blairi*, and *Rana chiricahuensis*.

In the plant world, Colin Bower has done something similar in his studies of *Chiloglottis* orchids. In some parts of New South Wales, Bower had the feeling that the *Chiloglottis* that grew in these areas was composed of several cryptic species, that is, species that were so similar in appearance that they could not be distinguished with complete certainty by visual examination of their flowers. He therefore cut specimens from what he thought might be the different species, popped them in jars, and moved them to various new locations, where he put them out in a line with each specimen separated from its neighbors by a foot or so. He then collected every wasp that came sailing in to inspect each orchid. The collective result of one set of experiments revealed that a certain population was indeed made up of three species, now named *Chiloglottis trilabra*, *Chiloglottis seminuda*, and *Chiloglottis reflexa*.

Bower reached this conclusion on the grounds that no matter where he put the three barely different-looking types of orchids, each one always attracted one and only one kind of wasp. Thus, the orchids that are now known as *Chiloglottis trilabra* were visited only by the wasp *Neozeleboria proxima*, whereas *Chiloglottis seminuda* had a different pollinator, albeit a wasp species in the same genus. Interestingly, males of this wasp were so similar in appearance to their close relative that, prior to Bower's work, entomologists had not detected the subtle but real differences in the external features of the two species. Likewise, the wasp that pounced on the flowers of a third species of *Chiloglottis* was yet another species of *Neozeleboria* that had not been named previously. So what we have here are three cryptic species of orchids pollinated by three cryptic species of wasps. Even if these orchids are capable of producing fertile hybrid offspring with their close relatives, as is true for many orchids, the wasps ensure that they do not have the opportunity to do so, keeping their gene pools separate.

The same thing happens with European orchids in the genus *Ophrys*, a group that contains many species that appear almost identical, but because they are pollinated by different species of bees in the genus *Andrena*, they are reproductively and genetically isolated from one another. Thus, for example, the physically similar *Ophrys fusca* and *Ophrys bilunulata* occur together on the Mediterranean island of Majorca, but the two orchids produce subtly different scent bouquets that result in the attraction of the two different pollinators. As a result, one orchid is pollinated exclusively by the bee *Andrena nigroaenea* and the other by *Andrena flavipes*. With no shared pollinators to promote genetic exchange between them, they can be considered separate species. Interestingly, the bee *Andrena nigroaenea* does pollinate another species of *Ophrys* as well as *O. fusca*, but because these two orchids deposit their pollinia on different parts of the bee's body (its head versus its abdomen), pollen from one is never applied to the stigma of the other. Because the members of the two populations do not reproduce together, they remain genetically distinctive units, that is, separate species.

The main point here is that just as differences in outward appearance do not guarantee that one is dealing with different species, so, too, similarities in size, shape, or color pattern do not prove that these similar-looking individuals belong to the same species. For many biologists, the trick is to determine whether groups of individuals are able and willing to reproduce, to exchange genetic material, to form an interbreeding aggregate genetically distinct from other populations. So what about the different flying duck orchids? Yes, some populations are slightly different in appearance from other populations. But are these physical differences based on the reproductive incompatibility of the different populations?

The answer is yes, probably. Remember that the key features that discriminate one flying duck from another reside in the labellum, especially with respect to the quantity and location of the glandular calli. The calli almost certainly play a critical role in the reproductive isolation of populations of flying ducks for the following reason. In these orchids, as in the hammer orchids and others, the glandular calli on the labellum probably get the attention of male thynnine wasps on patrol for mates of their own species. The calli apparently produce scents that resemble those of receptive female wasps. Males that pick up these odors may be lured toward the orchid; when close by, the wasp may then see the labellum, which acts as a visual decoy, stimulating the male to pounce on the petal as if it were a female wasp. Males that grasp the decoy, activate the flower's behavioral mechanism; as noted earlier, the curved hinge holding the labellum upright loses rigidity instantly, driving the labellum and the mounted wasp rapidly down into the cupped petals and sepals below, where the now sexually deflated, upside-down wasp may pick up or transfer a packet of pollen, thereby effecting pollination.

If populations of flying ducks have different patterns of calli (and differently shaped decoys), odds are that they are attracting different thynnine wasp pollinators. Andrew Brown has confirmed this point in at least some cases. Therefore, these pollinator-specific populations cannot exchange pollen and so are reproductively isolated from one another. In other words, they are different species—if one accepts the definition of a species as a population that fails to reproduce with the members of other populations because of one or another barrier to the exchange of genes.

Similar experiments are also needed to give us greater confidence about some of the other "species" of orchids that appear in Hoffman and Brown's field guide. Indeed, one of the major challenges of orchid identification in southwestern Australia is feeling confident that you have put the right name on the orchids you have located. The field guide, for all its merits, offers only one photograph and a few comments on key characteristics that users may refer to when attempting to identify a specimen in the field. Nothing more is needed for those species that look unlike any other, as a good many do, but an equally large contingent look devilishly alike to amateur botanists, if I am at all representative of this group of orchid hunters. Nor is it a great help to be told, as one sometimes is by Hoffman and Brown, that such and such a species can be "easily" separated from several other very similar ones on the basis of such characteristics as the relative size of their flowers, with one species possessed of larger (or smaller) petals, sepals, or labellum than some others. Perhaps if one had all the species in view simultaneously, one could reach a comfortable decision about which is what, but this luxury rarely presents itself. Moreover, many orchid species vary considerably in the size and color of their flowers even within a single location, so that separating species X from Y may require focusing on "average" or "typical" individuals, which may or may not be available to an observer. (Happily, Brown and his colleagues are working on an expanded field guide to the orchids of Western Australia that will feature multiple illustrations encompassing the variation within a species, which will be a great pleasure to have on an orchid-hunting expedition.)

Given that it is not at all obvious how to identify certain orchids, on what grounds did Hopper and Brown create so many species by dividing up what were once considered widespread but variable species? As an example of Hopper and Brown's work, consider their conclusions with respect to what was once called *Caladenia filamentosa*, a medium-size, thin-petalled spider orchid whose different populations had red or pink or white or pale yellow or bright yellow flowers with petals and sepals that were approximately 4 or 10 or 16 centimeters (nearly 6 inches) long. When Hopper and Brown revisited this "species," they declared it an entire complex of orchids, using a variety of features, including color, odor, and season of flowering, to differentiate among the multiple species they believed had been wrongly lumped together. They named one member of this complex *Caladenia luteola* partly on the basis of its bright yellow flowers. They pulled out another, somewhat more widespread subset of *Caladenia filamentosa* with *pale* yellow flowers and reinstated the name given to it by Lindley long ago, *Caladenia denticulata*. In their opinion, the differences in flower color and shape (coupled with an assortment of other features) were large enough to suggest that these two populations were genetically and reproductively isolated from each other and from other similar spider orchids, including those individuals that are still assigned the name *Caladenia filamentosa*, a species now restricted to Tasmania. So Brown and Hopper have exhibited a naming philosophy very different from that of their Australian colleagues, Burke and Adams, who collapsed *Dendrobium curvicaule*, *D. pedunculatum*, *D. rex*, and *D. tarberi* into *D. speciosum*, despite the differences in flower color and shape associated with different populations of this orchid.

The splitting of existing species into a complex of new ones is not unique to Hopper and Brown. The European orchid genus *Ophrys* was once believed to contain twenty-eight species (in 1929), a figure that has now reached eighty or so, according to some, or more than 140, according to others. Although some of the new species added since 1929 represent populations whose existence was not known then, many new species of *Ophrys* have been named by dividing a single previously described species into two or more separate entities. The justification for the splitting generally comes from the recognition of subtle structural differences previously overlooked or the acquisition of new information about pollinators that enables the taxonomist to differentiate between populations once lumped together under the same species name.

The case for reproductive isolation between two groups of plants (or animals, for that matter) is strengthened when evidence exists, as it sometimes does, that the two types coexist in the same area without crossing, as shown by an absence of intermediates of the sort that would be produced if members of the two populations interbred freely. Sometimes,

Orchids, Species, and Names

however, hybrids do occur at moderate frequency, and yet the parental species retain sufficient distinctiveness as populations to retain their very own labels. So, for example, the Margaret River spider orchid *Caladenia citrina* and the diamond spider orchid *Caladenia rhomboidiformis* look so different that the two species have not been merged despite the fact that the two hybridize fairly commonly. Indeed, after hours of searching for the Margaret River spider orchid on a special trip to the small area where it was reputed to grow, I only managed to find a strange specimen that had elements of what I was looking for combined with colors that clearly did not belong to the Margaret River spider orchid. The individual before me had the long yellowish petals of a Margaret River spider

5.3
Hybridization in the spider orchids. The two probable parental species (*Caladenia citrina*, far left, and *Caladenia longiclavata*, middle), may not look alike to us, but a male thynnine wasp visited both species, perhaps seeking nectar from *Caladenia citrina* and a sexual partner from *Caladenia longiclavata*. The hybrid offspring (left) exhibits a mix of characters from the two parents.

orchid but the dark red outer lip of the labellum of another species of spider orchid (probably *Caladenia longiclavata*). I had found a hybrid, something of a disappointment in that I could not claim to have added the uniformly pale yellow Margaret River spider orchid to my list of orchid species seen. But the disappointment was an interesting one nonetheless. Clearly, the two parental species were similar enough *genetically* so that the union of an egg cell from one species with a sperm cell from the other did not prevent the development of a fully equipped flowering plant. I wonder, however, if the hybrid was as likely as a "purebred" orchid to attract a pollinator and so pass on its mixed set of genes. If hybrids have difficulty reproducing, perhaps because pollinators often fail to recognize their mixed or intermediate attractant signals, they might fail to contribute to the eventual genetic merging of the two species.

Natural hybrids between most named orchids are rare, providing evidence that the different labels given to these species are valid. The scarcity of orchid hybrids stems in part because, as you know, many good species have their own special pollinator, a particular wasp, bee, fly, or what have you, and these specialists make relatively few mistakes. So when my wife announced that she had found a lovely sun orchid of a sort she had not seen before, she and I both assumed that the plant must be an uncommon species that we had not encountered until then. After hearing her description of the orchid's flower, which was magenta with darker spots, I thought that Sue's find was probably a specimen of the swamp sun orchid, one of the many species on my still-to-discover list. I encouraged Sue to take me back at once to the orchid in question, having on a few other occasions been tantalized by the description of an orchid of note that Sue was then unable to relocate. (In fairness to my spouse, many orchid hunters in Western Australia, myself included, have had the misfortune to think that we knew where we left an orchid, only to learn that we had vastly overrated our mental mapmaking capacity.) This time Sue marched me directly to the orchid in question so that I could take a close look, which I did after first apologizing for having doubted her ability to find the specimen again.

Now that the orchid was in front of me, I was sure that we had never seen anything like it before. Out came the orchid field guide, and soon I deduced that we were not looking at the swamp sun orchid; that orchid does have some dark red-purple markings on the flowers, but these are applied to a pale green background. Our orchid (note my shameless attempt to take partial possession of the orchid from my wife) had flowers that were primarily pale red-purple in color, not pale green. I went through the sun orchid section of the field guide two or three times, but to no avail. For a brief period, I let myself entertain the notion that we, or should I say Sue, had had the extraordinary good fortune to find a species new to science. The thought thrilled me and I was reluctant to let it go. On the other hand, I couldn't help but notice that the orchid's flowers combined features of two locally abundant species, the leopard orchid with its deep yellow, darker blotched flowers and a blue sun orchid with its uniformly radiant blue flowers. Moreover, as we crisscrossed the area surrounding the novel specimen, we failed to find even one other sun orchid that shared the look of *Thelymitra susanniana*, as I imagined it might be called by a professional botanist, who would, of course, name the new orchid in honor of its discoverer. Sadly, there was no getting around it: we had a hybrid here, admittedly a beautiful, rare, and distinctive hybrid, but not a previously undiscovered species. Later, Andrew Brown confirmed this diagnosis.

Given that two structurally different species only very rarely combine male and female gametes to create a mistake, a hybrid, a novel creation, one wonders if those rare mistakes mostly involve two species that happen to be genetically fairly close to one another. One would predict that the scents produced by such pairs of orchids would be more similar than the scents released by distantly related pairs. If true, the pollinators of species A might become confused once in a while, producing a hybrid when they visited first species A, then species B. Although in a greenhouse, horticulturalists can manually create flowering hybrids from parental species that belong to very different groups, these kinds of crosses are exceptionally rare when the pollinator is a wasp or bee or fly rather than a pollen-transferring human being.

The point is that there are good species out there in the universe of orchids, populations that are essentially incapable of exchanging genetic material with other populations, often because each group has an evolved dependence on their own highly specific pollinator. But not every population can be unequivocally assigned to a particular species. Some groups

5.4
Hybridization also occurs in the sun orchids. The two probable parental species (right and middle), and the product of a cross between them (far right).

Orchids, Species, and Names

are only partly separated reproductively, and indeed may never break completely away from one another. Other distinctive populations are geographically separated, but are they really reproductively separate? In other words, are the differences in the appearance of the plants from two locations great enough to ensure that they would not exchange pollen *if* they were to coexist in the same area? One could in theory get a handle on this question via molecular genetic comparisons or, as noted earlier, by collecting the pollinators visiting flowers in the two areas. If the populations were highly similar genetically and if they were both pollinated by the same insect species, then one could make a convincing case that both populations belonged to the same species. If, however, the populations were genetically distinctive and pollinated by different insects, then we

would be inclined to judge them separate species. But in practice, securing these data is so challenging that plant taxonomists often make judgments based strictly on the appearance of a particular population. Here intuition can play such a dominant role in the final decision as to make the rest of us a bit uncomfortable.

Take the big white spider orchid, *Caladenia longicauda*, a species that exhibits considerable size variation within any one of its numerous and widely distributed populations. Hopper and Brown are fully aware of this variation but believe that one can nevertheless recognize subtle average differences between plants that they place within *longicauda* and other large, long-petalled white spider orchids, which they have put in new species, such as the white sandplain spider orchid, *Caladenia speciosa*. I have peered long and hard at the field guide and even have read the scientific description of the species in Hopper and Brown's monograph, in which they carefully lay out the basis for separating *Caladenia longicauda* from *Caladenia speciosa*. I am still quite unable to say that I would recognize *Caladenia speciosa* if I had one right in front of me, so similar are these two species and so great is the variation within the common *Caladenia longicauda*. Nonetheless, I have no doubt that Hopper and Brown could make these discriminations based on their vastly greater experience with the Western Australian orchids.

On the other hand, some differences between superficially similar populations of Australian orchids are so great that even a novice orchidophile has little trouble detecting them. I know that I, for example, had no trouble distinguishing the common white spider orchid from another white-flowered spider orchid found on Mt. Ridley. Although Mt. Ridley is less than 100 kilometers from the coastal town of Esperance and its surrounding wheat fields and pastures, it might just as well be in a different universe. No cleared fields, no human habitation, no other cars, just the low spindly eucalypts and spiny shrubs surrounding the small circular salt lakes in a vast wilderness. When we last visited Mt. Ridley, the improved dirt road ended abruptly some five miles or so from our destination. We paused to survey the ominously rutted white-clay surface of a track that ran the rest of the way to the dark gray mountain of granite that we wished to inspect. This definitely unimproved roadway was not one to inspire confidence in a driver. I would never have been tempted to experiment with it on a wet day, but as it had not rained recently, I decided to risk it. We encountered no difficulty in following the track as it weaved unsteadily through the woodlands, until it came to the rock itself, one of those glorious reserves in Western Australia that only naturalists and occasional bush campers are likely to visit. When we arrived, a wedge-tailed eagle with its great wings held up in a triumphant V drifted slowly in giant circles over Mt. Ridley, which presided over an unbroken wilderness of dry woodland and salt lakes to the north and east.

I had not come to Mt. Ridley for its spider orchids but to find a rufous greenhood orchid that the field guide told me grew only on a few granite outcrops in this part of Western Australia. After parking in a glade by some taller-than-average eucalypts, we admired the wedge-tailed eagle for a while before beginning the hunt. Sue and I climbed slowly up the stony face of the rounded Mt. Ridley and were treated as usual to some initial disappointments. At first, we failed to find anything at all except, surprisingly, the rather ugly nonnative South African orchid *Disa bracteata*, which was common in the soil pockets that had formed in shallow depressions gouged out of the granite. Leaving these undesirables behind, we then found some old, partly withered specimens of what might have been the very greenhood I had hoped to find in flower. Embedded in this discovery was some good news and some bad news. The orchid was probably here as promised, which was good. But perhaps we had missed its flowering season, which would have been unfortunate.

With evening some hours away, we had the time and desire to find rufous greenhoods with intact flowers. As we scanned the rock surfaces, a little party of thornbills came along, flitting through the droopy pine-like casuarinas that had colonized a deep trench in the rock near its top. Shortly thereafter, I nearly stepped on the species that we were after. A fresh specimen barely three inches tall had somehow found sufficient soil in a minute crack in the granite. Despite its surroundings, the orchid was doing fine, as befits a member of a species that specializes in overcoming the hardships provided by granite rock habitats. The Mt. Ridley greenhood is a currently undescribed species, although it is about to be named by a botanist who will justify giving the individuals growing on the granite outcrop their own species' name on the reasonable grounds that this population and several others found on nearby granite rocks all share some subtle physical characteristics found in no other populations. Moreover, these orchids also have a different flowering season and a geographic distribution distinct from that of other similar-looking, but slightly different, rufous greenhoods.

5.5

The Mt. Ridley rufous greenhood, an undescribed species that is restricted to a few granite outcrops in a small part of southwestern Australia.

These differences should translate into genetic differences between the Mt. Ridley rufous greenhood and very similar orchids that flower earlier in other parts of Western Australia. This is a potentially testable proposition, thanks to the various molecular genetic techniques that taxonomists now use. In fact, one such technique has already been used to quantify the genetic differences between six similar-looking Western Australian members of the genus *Pterostylis* that had been given specific labels on the basis of small structural differences among them. A research team located fifteen variable genes present in the six "species." The team then established the frequency with which a given variant form of a given gene appeared in every population sampled for all six species. With the numbers generated in this exercise, the orchid crew could complete an equation that produces a measure called *genetic distance*, a single figure that quantifies the genetic difference between two populations. As expected, genetic distances were very small for populations of orchids that had been placed within the same species. These groups had presumably regularly exchanged genes with one another. As was also expected, the genetic distance values were higher, often much higher, for populations previously assigned to different species, as in *Pterostylis scabra* versus *Pterostylis rogersii*, which are both genetically and geographically separated. This investigation therefore confirmed that all the populations considered to be a particular species, such as *Pterostylis scabra*, formed a cluster distinct from other groups belonging to the other species, such as *Pterostylis rogersii*. The fact that genetic information backed up a classification scheme that had originally been developed on the basis of nonmolecular comparisons gives us considerable confidence that the species' designations are valid.

The examination of a set of variable genes shared by a set of putative species is only one of several techniques available for comparing the genetic constitution of different groups of organisms. Unfortunately, no agreement exists on exactly what degree of genetic distance is required before two populations are to be considered definitively different species. In theory, a single gene might affect something like the production of a chemical vital to pollinator attraction, so that two populations that differed with respect to just this one gene might attract different pollinators. In such a case, a very small genetic difference could possibly result in complete reproductive isolation between the two groups of orchids. At the other extreme, large numbers of genetic changes might occur in two geographically separated groups without affecting the reproductive systems of the individuals belonging to these groups, so that they might well still be able to interbreed successfully, if given the chance.

The bottom line is that no single fixed criterion exists for resolving whether population X belongs to species Y or should instead be considered a separate species in its own right. Therefore, disagreements and uncertainties in classification are to be expected. And they occur. For example, are the yellow lady's slippers of North America (see Figure 1.7) to be classified with those from Europe and Asia as *Cypripedium calceolus*, or are they to be considered a different species that deserves its own name, *Cypripedium parviflorum*? Are the North American yellow lady's slippers with unusually large flowers that occur in some parts of the southeastern United States different enough to warrant separation as the species *Cypripedium kentuckiense*? These populations do have larger flowers than other yellow lady's slipper populations, but the common yellow lady's slipper is known to be highly variable with respect to the color and size of its flowers. The plants that have been labeled by some as *Cypripedium kentuckiense* do have a unique form of one particular gene, but they have the same forms of eleven other genes that occur in populations of *C. parviflorum*. I would say that there is no clear-cut answer here to the question of species identity.

I had the opportunity to come to grips with uncertainties of this sort on Mt. Ridley, which I did after triumphantly finding and photographing the Mt. Ridley rufous greenhood. Because we still had time before having to return to the van to prepare dinner, we continued wandering around on Mt. Ridley. Having reached the peak itself, only a few hundred feet above the almost perfectly flat surrounding plain, we began to zigzag down the rock. Mt. Burdett was clearly visible far out to the east, a companion granite bump on the horizon now illuminated wistfully by the descending sun. As I angled along a barren granite avenue between two patches of low shrubs that had managed to grab toeholds on Mt. Ridley, I glimpsed something beneath a shrub, a flash of white that turned my head: under the shrub, a cluster of spider orchids. Unfortunately, all except one had flowers with folded and faded petals and sepals, but the one remaining intact specimen, even near the end of its game, was a charming little thing, a miniaturized white spider orchid, a Chihuahua of a plant compared to the greyhounds that I had seen earlier on the trip far to the west. Despite its small size and its short and rigid petals so unlike the great drooping petals of the large white spider orchids, the gestalt of the labellum nevertheless reminded me of the *Caladenia longicauda* with which I had become familiar.

Sure enough, perusal of the field guide revealed that I was probably looking at a subspecies of *Caladenia longicauda* called *Caladenia longicauda rigidula*. Now a subspecies is not a full species, merely a population judged by a professional taxonomist to be sufficiently distinctive from other geographically separate populations of the same species to warrant special recognition via the addition of a third name to its scientific moniker. A subspecies can conceivably be a population on the way to becoming a new species—eventually. But no guarantee exists that a subspecies will ever acquire the requisite number of genetic changes needed to block its reproduction with other populations of the parental species. Thus, a subspecies is simply a geographically separate population whose special appearance, genetics, or reproductive system is different from certain from other populations, but not so different as to warrant giving the subspecies a unique species name of its own.

You may have noticed that I said I was "probably" looking at the subspecies *rigidula* of *Caladenia longicauda*. I make this disclaimer because the field guide contains a photograph of a small white spider orchid called *Caladenia cruscula*, which to me looks identical to *Caladenia longicauda rigidula*. As best I can tell, the only substantial difference between them is one of habitat, with *Caladenia cruscula* said to be fond of the borders of the salt lakes that occur so commonly in the general vicinity of Mt. Ridley. Because I found my miniature white spider orchid on the slopes of a granite rock rather than by a salt lake, I believe I can rule out *cruscula* as its name.

5.6

Variation within the species *Caladenia longicauda*. The specimens in the top photo grow near Esperance, Western Australia, and have large flowers with great dangling petals and sepals; they belong to the subspecies *crassa*. In the bottom photo is a representative of a population found only 50 kilometers north of Esperance but that are only a quarter of the size of those on the top; it belongs to the subspecies *rigidula*.

The point is that I lack sufficient experience with the *Caladenia longicauda* complex of species and subspecies, which creates problems for me when confronted with one of the more tricky members of the group. Those with more experience with these orchids and an eye for these matters can, I imagine, develop the abilities needed to sort through individuals and assign them confidently to one group or another. But because intuitions can differ, and evidence can vary, plenty of room exists for debate not just about whether population X is part of subspecies Z or species Y, but also even about which species belong in genus A versus genus B.

For example, all the flying duck species in Western Australia have been placed in the genus *Paracaleana* by Hoffman and Brown, a deci-

185

sion accepted by many of their botanical colleagues. The name of this genus was created in 1972 by Donald Blaxell, yet another Australian botanist, when he changed the original scientific label of the common flying duck orchid from *Caleana nigrita* to *Paracaleana nigrita*. Blaxell did so to emphasize what he thought were important structural differences between what were then called *Caleana nigrita* and *Caleana minor* versus a third species, *Caleana major*. The genus *Caleana* was the invention of Robert Brown, who in 1810 honored his contemporary George Caley, an orchid collector in Australia who had supplied Brown with many interesting species to describe. When the common flying duck orchid was first named by John Lindley in 1840, he felt it belonged in Brown's genus *Caleana*. Blaxell disagreed for a variety of technical reasons, and he said so in an article in the *Contributions from the New South Wales National Herbarium*, which followed established procedures for the scientific renaming of species. Although Hopper and Brown agreed with Blaxell, others initially persisted in keeping the western flying ducks in the genus *Caleana* with *Caleana major*. At the moment, most of those who had once urged that *Paracaleana* be dropped and that all the eastern and western species of flying ducks be bundled into *Caleana* have changed their mind on the basis of genetic comparisons among the various species. But there is no absolute guarantee that minds won't change again in the future.

5.7

Two species of flying duck orchids that occur in eastern Australia, which at one time were both placed in the genus *Caleana*: the tiny *Paracaleana minor* (left) is much smaller than *Caleana major* (right).

As an outsider to the taxonomic tussles in the orchid world, my opinions on orchid classification are not in great demand, but for what it is worth, I happen to think that it makes sense to leaves the eastern flying ducks called *Caleana major* in the genus *Caleana* and put the rest into the genus *Paracaleana*. For one thing, most of the Western Australian species have extremely similar flowers whose shape and calli differ consistently and substantially from the flower shape and calli distribution of their eastern Australian relative *Caleana major*. Those differences that do exist among the species that currently belong to *Paracaleana* are relatively modest. Although some would divide these species into two subgenera, which is the category just below the genus level, no one has suggested creating two separate genera for the species now situated in *Paracaleana*. But as this case illustrates, an element of subjectivity comes into play in the world of classification, which is why differences of opinion and interpretation occur about what goes in this genus or that, or whether a particular population deserves its own species name or not.

So it is not just *Caleana* versus *Paracaleana*. Similar arguments have, for example, gone back and forth about the spider orchids and their relatives. Botanists have long wondered whether the little dragon orchids of Western Australia should be classified with the spider orchids (genus *Caladenia*) or placed in their own genus in recognition of their special anatomical features, most notably the weird tick-like female decoy possessed by the four dragon orchid species. The early workers concluded that, despite the novel nature of the dragon orchids' labellum, the gestalt of the flower was close enough to that of some spider orchids to justify their placement in the genus *Caladenia*. However, when Steve Hopper and Andrew Brown revisited this decision in 1992, they felt that the structural differences between dragon and spider orchids were so great that the dragon orchids should be given a generic name of their own, which they informally supplied as *Drakonorchis*.

Hopper and Brown's claim at that time can be interpreted as a hypothesis about the evolutionary origins of the dragon orchids, namely, that they had split off quite some time ago from the *Caladenia* lineage. If this hypothesis is correct, the putative distant common ancestor of the modern dragon and spider orchids must have given rise to two branches, with one line eventually evolving into the modern dragon orchids and the other line going its own evolutionary way to become the modern spider orchids. The substantial period of separation should have resulted in the accumulation of many genetic differences between the two lineages. As you know, this prediction can be tested now that molecular analyses are more or less routine. The data are in, collected by a large research team led by Paul Kores (and including Hopper and Brown). The results, published in 2001, show that contrary to Hopper and Brown's original view, the dragon orchids are embedded within the large group of orchids that at the time were all considered members of the genus *Caladenia*. Indeed, some members of *Caladenia* differ more from one another in molecular

5.8
Should the dragon orchid (top) be placed in the same genus as this spider orchid (bottom)? The molecular data say that they are close relatives.

terms than they do from the dragon orchids. These findings properly led Hopper and Brown to leave the dragon orchid in its original genus; the species that they had informally named *Drakonorchis barbarossa* returned to its old name, *Caladenia barbarossa*.

This decision, however, has not gone unchallenged. The Polish botanist D. L. Szlachetko published a revision in 2001 in which he broke up *Caladenia* into several genera, each with its own smaller stable of species. This new system did not satisfy a group of Australian botanists composed of David Jones and his colleagues at the Commonwealth Scientific and Industrial Research Organization (CSIRO). Jones and company have a long record of classifying Australian orchids, and so they bring plenty of experience to the taxonomic table. But although they disliked Szlachetko's revision, they also disagreed with Hopper and Brown, in keeping with the tradition of taxonomists who can be moderately contentious on occasion. In a paper published in *The Orchadian*, Jones and company take a hatchet to the genus *Caladenia*, splitting it up in ways different from Szlachetko's scheme, but fractionating it nonetheless.

Their new classification is based on an analysis of the information provided by specific segments of DNA taken from thirty-two representative species of the "old" genus *Caladenia*. Genetic data are obviously highly useful in this regard, particularly when, as is true here, researchers use base sequence information from sections of DNA that do *not* code for particular proteins. Noncoding DNA makes up a large proportion of most genomes for reasons that are still not clear to biologists. Because these stretches of the molecule have no apparent function, random changes that occur by mutation in these regions can persist indefinitely. Why? Because they are hidden from selection, thanks to their inability to influence development. We can assume that over time, more and more hidden or "neutral" changes will build up between species descended from a common ancestor. In other words, a simple count of the number of base differences between two species provides a relative measure of how long they have been separated.

The CSIRO genetic study revealed six clusters of species distinguished on the basis of their DNA. The members of each cluster share similar base sequences that differ from those of the other groups. Because of the molecular data, Jones and company gave each of the six groups its own generic label. One of their genera is *Drakonorchis*, which they resurrect, overruling Hopper and Brown's decision to merge the dragon orchids back into the genus *Caladenia*.

Hopper and Brown accept the validity of the molecular comparisons presented by the CSIRO team, but they disagree on where to draw the lines between genera. If, like Hopper and Brown, you wish to stress the fact that all thirty-two species from *cardiochila* to *lyallii* share a common ancestor (basal species A on Figure 5.9), one clear way to do so is to put all thirty-two species in the same genus (*Caladenia*). Perhaps, however, you, like the CSIRO team, wish to draw attention to the fact that *congesta*, *praecox*, and *lyallii* constitute one species cluster with ancestral species G, a more recent ancestor than A and different from other more or less contemporaneous ancestral species B, C, D, E, and F, each of which gave rise to its own cluster of living species. If so, you will place the descendants of each of the ancestral species, B through G, in their own genus. The CSIRO researchers retained the generic name *Caladenia*, but they applied it only to the cowslip orchid, *Caladenia flava* (see Figure 5.9), and a few others (the fairy orchids) that also have short, nonspidery petals and sepals.

Because there is no rule that says a genus absolutely must be based on such and such criteria, the three systems of Hopper and Brown, Szlachetko, and Jones's team are out there competing with one another. General acceptance of one and rejection of the others will depend in part on whether the highly specific and sometimes arcane rules of classification have been properly followed. In this regard, Hopper and Brown have recently identified a number of violations of the rules in the work of the other researchers that clearly require correction (and consequently, some renaming). Perhaps more important, they point out that it is rare for experts in classification to take one established genus that really is composed of closely related species and break it up into a larger number of genera. The name *Caladenia* has been around for nearly two hundred years and botanists are familiar with it and what it stands for as a statement about the evolutionary relationships among the 243 species that Hopper and Brown believe the genus contains. To chop this group into six subgroups, each with its own generic label, requires, if nothing else, that persons interested in these orchids learn five new names. I know where I stand on this matter; for me, the fewer new names to learn, the better. As Hopper and Brown point, there is much to be said for something called *nomenclatural stability*, the use of well-established names, provided they do not violate what we know about the evolutionary history of the species included in the taxonomic groups so named. But it will be up to the professionals to decide this matter more or less by consensus, something that will not happen immediately and will probably depend in large measure on the research philosophies and intuitions of the academics involved.

Genus, species, subspecies—does it really matter all that much what goes into which slot? Actually, I think it does, and for two quite different reasons. First, classification schemes can be hypotheses about the evolutionary relationships

5.9

A molecular comparison of thirty-two orchid species used by two different research groups to produce two different classification systems. In system 1, all the descendants of ancestral species A are placed in the same genus (*Caladenia*); in the alternative system 2, the descendants of ancestral species B to G are separated into genera of their own. (For technical reasons, the genus *Glychorchis* is not now considered valid by systematists; instead, it should be *Ericksonella*.)

among species. Therefore, an "incorrect" classification is one that makes an incorrect claim about which groups of species share a recent ancestral species and which do not. So a desire for truth in advertising requires that the classifiers do the best they can at getting the relationships right.

In this regard, the various kinds of molecular data are generally considered more likely than any other information to yield an accurate

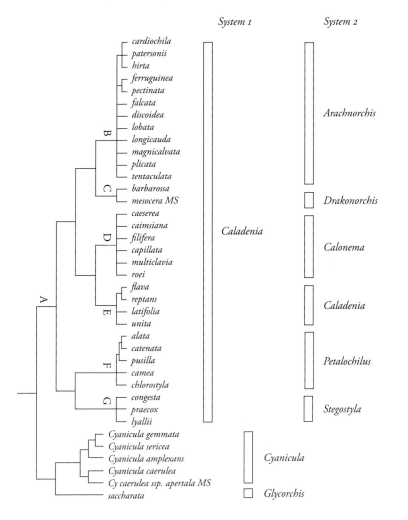

depiction of evolutionary relationships. Most of the early orchid classification schemes were built on a handful of structural characteristics (generally, anatomical features of the flowers) that orchid taxonomists intuitively felt were of exceptional importance. In contrast, molecular genetic analyses embrace a vastly larger set of characteristics: the bases in long sequences of nucleotides. As a result of being able to work with more characters, taxonomists can produce more robust descriptions of the similarities and differences among species, which is especially useful when testing evolutionary hypotheses built with different, and fewer, characteristics.

Let me note, however, that molecular studies do not eliminate debate and disagreement. Molecular studies depend on which protein or chunk of DNA is available for comparison. Conclusions based on some molecules do not necessarily match those based on another biochemical. Thus, the interpretation of molecular data can be downright contentious. Nor should you think that once agreement has been reached among molecular taxonomists, their conclusions always improve those based on "old-fashioned" techniques. Indeed, molecular and morphological data regularly point to the same system of classification. This is cause for celebration because it suggests that a particular scheme really can be trusted. So, for example, in 1960 most of the roughly 850 genera in the family Orchidaceae were placed by Robert Dressler and Calaway Dodson into one of five major groups largely on the basis of the number of stamens and other aspects of flower structure. In effect, Dressler and Dodson were arguing that these five groups, or subfamilies, represented distinct evolutionary lineages that could be traced back to five different ancestral orchid species. When a large research team of molecular botanists reexamined this scenario using nucleotide sequence comparisons of a particular gene found in nearly every orchid tested to date, they discovered that the Dressler-Dodson five-part scheme largely matched the results that came from their DNA analyses (although debate continues on just how many subfamilies should be recognized). When molecular and anatomical comparisons yield much the same conclusions, one can have more confidence in the validity of the evolutionary relationships they are intended to describe.

The second reason classification matters, in addition to the academic importance of accurately describing evolutionary relationships, has to do with the conservation of biodiversity. Just what does one conserve? People generally treat an entire species as more important than a subspecies or genetically differentiated population when it comes to establishing what gets protected. If the main goal of a conservation program is to preserve threatened species, it makes a difference whether endangered population X is considered a species in its own right or is instead "merely" one of many populations that belong to a common, widespread and variable species. One *could* make the case that a distinctive, strikingly colored population of widespread species Z deserves protection in its own right, but selling that argument may be harder than getting help to preserve a rare and endangered species of limited distribution. To the extent that genetic studies lead to well-supported classification systems, conservationists can make use of them to assert, perhaps persuasively, that a particular endangered population is indeed a genuine species and thus especially deserving of protection. (The Western Australian legal system does not distinguish between species, subspecies, or distinctive populations, granting all equal protection under the law. But even here I suspect that there may be a greater sense of urgency in establishing that a species, as opposed to a subspecies, is threatened and deserving of active protection.)

There is another way in which robust classifications can have conservation utility. Given the sad reality that the money available for plant rescues and reserve formation will always be in short supply, establishing conservation priorities is probably necessary. Some persons have argued, therefore, that in a triage situation, evolutionary novelties should receive top priority. So, for example, given the cruel choice of investing limited resources in the protection of an endangered eucalytpus, one species among hundreds, as opposed to the Albany pitcher plant, a species with no especially close relatives, perhaps the nod should go to the Albany pitcher plant. If such a rule of thumb became conservation policy, then an accurate demonstration of the position of a species on an evolutionary tree would become critical. Which of two species in competition for the same conservation dollar is more distinctive, genetically and evolutionarily speaking? Answering this question will surely require molecular genetic studies to ensure that policy decisions are based on scientific conclusions that are as strong as possible.

It is testimony to the abundance of species, the relative recency of molecular taxonomy, and the shortage of modern botanists willing and able to classify plants that so much work remains to be done on the naming front. The two flying duck orchids of Cape Arid have not yet officially received their own proper scientific labels, nor have their genetics been analyzed for similarities with other Western Australian flying ducks, to say nothing of the flying duck orchid *Caleana major* of eastern Australia. Are the members of *Paracaleana* really molecularly distinctive from *Caleana major*? I can't wait to read the research that answers this question. Sure, once the work is done, no great medical breakthrough will follow, no new stock offering will go on the market, and no newspaper will broadcast the findings on the first (or even sixty-fourth) page. But I'll be interested and pleased that another tiny piece of the huge evolutionary puzzle of life has been put in the right place, giving those of us who care a slightly better understanding of the history of the plants we love.

Thanks to Steve Hopper and Andrew Brown's continuing taxonomic work on Western Australia's *Caladenia* orchids, those of us looking for spider, fairy, and dragon orchids now have than 107 species to track down if we accept their view that these species deserve placement in this one genus. The diversity of form within this group, however classified, makes these orchids highly appealing to amateurs and professional botanists alike. I have tried to find and photograph as many species as possible, and to date my list is up to sixty-two, including such relatively rare ones (I am proud to announce) as the giant spider orchid, *Caladenia excelsa*, a plant that can reach a meter in height with flowers that have great drooping petals and sepals, each one as long as your hand, and the limestone spider orchid, *Caladenia bicalliata*, a miniature species whose flowers are less than an inch across.

6

Orchids,

Biodiversity,

and Hotspots

My mission to the spider orchids was rewarded recently with the discovery of yet another species, the Walpole spider orchid, *Caladenia interjacens*, one of the many Western Australian species with an almost ridiculously limited distribution. The orchid only grows within a 30-kilometer coastal stretch between the little town of Walpole and West Cliff Point. When I began my Walpole spider orchid quest, my wife and I went first to the town itself. We did find a specimen there, in a little tourist center, which featured cut specimens of the local orchids. The Walpole spider orchid, its big white flower blushed with pink and ornamented with long dangling petals and handsome sepals, dominated the display, making me long to see it in something other than a water-filled jar next to a rack of cheap souvenirs and glossy tourist brochures. But when I asked the woman manning the tourist center desk for advice, she said that she had no idea where the orchid came from. She went on to add ominously, "I think it's too late in the season to find that species now."

We were not, however, about to give up without a fight, no matter how much our contact at the Walpole tourist center wanted to temper our enthusiasm with realism. Having not received outside help on where to look for the orchid, I decided to make the most of the field guide, where I remembered reading something about a West Cliff Point. I dug the road map out of its home in the campervan and spent some time peering along the coast to the *east* of Walpole, having vaguely remembered the orchidologist Bill Jackson telling me some years previously that the Walpole spider orchid was reasonably common around Conspicuous Cliffs. This landmark was prominent on my map, but I could see no mention of West Cliff Point.

As noted earlier, Gloria and the late Bill Jackson were devoted orchid hunters who made a habit of discovering new and rare species. They were one of the first to realize that the Walpole spider orchid differed from the other large white spider orchids of Western Australia and so deserved its own name, *Caladenia interjacens*, which the orchid formally received only in 2001. In that year, Hopper and Brown published their monograph in *Nuytsia*, a specialized botanical journal, where they described the key features of the Walpole spider orchid in the turgid manner required of professional taxonomists ("lateral lobes obliquely ascending with entire margins near the claw, becoming fimbriate with slender clubbed narrowly fusiform pale pink to white calli to 7 mm long which are abruptly decrescent near midlobe").

Suffice it to say that the formal description of the orchid fails to capture its aesthetic appeal. In my eagerness to see the plant rather than read about it, I decided to head out toward Conspicuous Cliffs, as West Cliff Point did not seem to be on the map. You may wonder why I did

6.1
The differences among even fairly closely related orchid species can be huge, as illustrated by the tiny flower of *Caladenia bicalliata* versus the long-petaled flower of *Caladenia excelsa*.

not phone Bill Jackson and enlist his assistance in finding the orchid. First, I had little desire to impose on someone whom I had met only once before. Second, I knew it would be more fun to find the orchid on my own, with the assistance, if needed, of my wife and our travel companions in Walpole. True, we ran the risk of dismal failure, but even a small possibility of experiencing the thrill of personal discovery was good enough for me.

In our effort to achieve that thrill, we spent the better part of three days walking the trails near Conspicuous Cliffs, where I thought we had a chance to find the orchid in question. Inland from the massive limestone cliffs, old sand dunes support everything from waist-high shrubs to skinny peppermint trees to massive karri eucalypts. The orchid guide states that the Walpole spider orchid occurs on consolidated sand dunes in areas with low open heath or peppermint trees, and so we had every reason to think we were on the right track. But although the trails were excellent and the habitat promising, we failed to find a single Walpole spider orchid.

Toward the end of this less than successful phase of our search, my wife took it upon herself to ask a ranger where we could locate the plant, whose photograph adorns the display case at the end of the road to Conspicuous Cliffs beach (giving us another reason to think our search had been on target). The ranger, however, professed ignorance of the orchid's whereabouts. Some members of our party became resigned to failure. Perhaps the orchids had come and gone, as the tourist center operative had suggested.

Still resolute, Sue picked up the orchid field guide and read again its brief description of the plant, its habitats and location. In so doing she wondered out loud where West Cliff Point was. Her musing caught the ear of Henry Bennett-Clark, a fellow biologist and one of our companions on this trip. Henry took it upon himself to look at a large map of the local surroundings that happened to be pinned to wall of the cabin in which we were staying. Unlike me, Henry had no preconceptions about where West Cliff Point might be, and so he looked both to the left and right of Walpole, finding that the peninsula in question was some 30 kilometers to the west of Walpole, not to the east, as I had originally and erroneously assumed. Because of yet another geographical slipup, I had us looking in the wrong place for the restrictively distributed Walpole spider orchid. I took the subsequent criticism of my map sense and orchid searching judgment in good grace, if I do say so.

Even with information about the location of West Cliff Point in hand, we had no guarantee that we would find the Walpole spider orchid, but at least we knew where *not* to look. Rather than spend more time peering at the ground near Conspicuous Cliffs, we traveled down a dirt road that went to Mandalay Beach, about halfway between Walpole and the now accurately located West Cliff Point. Within minutes of starting down the road, we found ourselves in an area of stabilized sand dunes covered with low heath and groves of peppermint trees. We stopped the car and all five of us spilled out to search the site. Despite the agreeable match between the habitat in which we were standing and the supposed requirements of the orchid, about half an hour passed before one of our company, my wife, who was walking down the road rather than wandering among heath and peppermint, called out for me to come. I knew from her triumphal tone of voice that she had succeeded in finding the orchid. The plant, which was growing within a few feet of the road, was every bit as handsome as promised. The living gem in front of us made the cut specimen in the tourist bureau seem almost tawdry by comparison. Congratulations flowed freely, albeit with an occasional comment from some on my now well-documented deficiencies as a reader of maps.

The Walpole spider orchid is not the only species of *Caladenia* that occupies a very tiny enclave in greater Western Australia. The somewhat

Orchids, Biodiversity, and Hotspots

smaller, but equally gorgeous, exotic spider orchid, *Caladenia nivalis*, also has a home range of just a few kilometers of coastal heath, and the populations of any number of other species are restricted to exceedingly small areas as well. The abundance of highly localized species is one reason the total number of Western Australian caladenias (broadly defined) tops the century mark, a stunning number particularly when you realize that California, a state similar in size to the southwestern portion of Western Australia, has only about thirty species of orchids of any sort to its name. Moreover, as you know, *Caladenia* is by no means the only species-rich genus of orchids in Western Australia, which is home to dozens of species in the genus *Pterostylis*, as well as many sun orchids (*Thelymitra*), a handful of hammer orchids (*Drakaea*), some flying ducks (*Paracaleana*), and the list goes on. The grand total for southwestern Australia: well over three hundred species, with still more to be catalogued.

6.2

The Walpole spider orchid, *Caladenia interjacens*, one of the many Western Australian species with a very limited distribution.

How can we account for the striking richness of the orchid flora of this part of the world? The first thing to be said is that the orchids are only one of several exceptionally species-rich plant groups in southwestern Australia. Here you will find, for example, an abundance of shrubs and small trees in the family Proteaceae, which embraces the wonderful banksias, dryandras, grevilleas, and hakeas with their large, colorful, and complex inflorescences. Of the seventy-five species of banksias that make Australia home, sixty (80 percent) occur only in the southwestern corner of the country, despite the fact that this region makes up only 5 percent of the total land mass of Australia. Of the roughly 170 Australian grevilleas, about half grow in the southwest, often within postage-stamp domains. There are perhaps ninety-two species of dryandras, and every one is found only in southwestern Australia. Outside of the Proteaceae,

Isopogon

Dryandra

204

6.3

Diversity in the Proteaceae
of southwestern Australia.
Representatives of four
species-rich genera belonging
to this family: an *Isopogon*,
a *Dryandra*, a *Grevillea*,
and a *Banksia*.

Grevillea

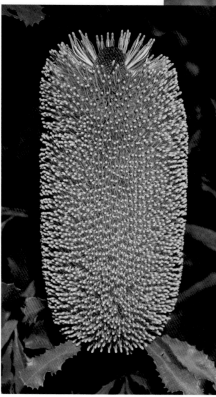

Banksia

some other groups of plants have also gone wild here. You say you would like to find some acacias? Well, let me recommend southwestern Australia, with its roughly five hundred species, as compared to a grand total of six in the whole of Arizona, my home state.

The dozens or hundreds of species in each genus indicate that an unusually high rate of speciation followed the establishment of an ancestral *Banksia, Dryandra, Grevillea, Hakea,* and *Acacia* in southwestern Australia. The cumulative effect of the splitting of a few lineages has been the evolution of a host of species belonging to just a handful of genera. The Proteaceae provide the region with only sixteen genera but about seven hundred species of plants, a substantial part of the grand total of seventy-five hundred. Overall, about 75 percent of the southwestern plant species occur only here, nowhere else in the world, indeed nowhere else in Australia, for that matter.

The count of seventy-five hundred plant species takes one's breath away when one considers that all of Great Britain holds only about sixteen hundred species and the state of New York just barely tops two thousand. In fact, there are more species on the threatened and poorly known list (more than twenty-two hundred species) for Western Australia than the totals for all flowering plant species in Great Britain or New York. The southwestern nose of Australia is a plant diversity hotspot, so much so that the region made it into the top twenty-five areas in the world in terms of numbers of unique plant species.

Just why should this part of Australia be so botanically diverse? To understand why, we need to recall that a species is generally considered to be a population that has become reproductively (and genetically) distinct from other populations. Therefore, speciation is the evolutionary process that produces a reproductively (and genetically) distinctive group of individuals. This process is usually thought to require a period of geographic isolation, during which subpopulation X undergoes a series of genetic changes without reproductive contact with other populations that in the past formed one freely interbreeding unit.

The geographic isolation of proto-species may have occurred often in southwestern Australia because of the many rapid climate changes that have taken place here during the past few million years. Currently, the southwestern nose of Australia is relatively cool and wet because cold fronts roll in off the Indian Ocean to make landfall here. Prior to extensive logging, this portion of the state was blanketed in a temperate rainforest of giant karri eucalypts, which can hold their own against California redwoods in the size sweepstakes, thank you very much. As one goes north or east from Cape Leeuwin, annual precipitation falls off quickly and the vegetation changes correspondingly. In the slightly drier zones, the forests are composed of somewhat smaller eucalypts (with some exceptions), which are then replaced by still smaller and more drought-tolerant species in the semiarid band that intervenes between the coast and the genuine inland deserts of southwestern Australia. Although there are some surprisingly large eucalypts in areas that receive less than ten inches of rain per year in western Australia, much of the true desert ecosystems are dominated by relatively small, extremely drought-tolerant acacias.

The zones of wet to dry have shifted considerably over the past several million years. When wetter and cooler conditions prevailed, the temperate zone was much more extensive, pushing the edge of the desert further into the interior; conversely, when the trend was toward warmer, more arid conditions, the edge of the desert has moved closer to the coast. Relatively rapid changes of this sort could have fragmented previously widespread species into an archipelago of isolated populations, some of which occupied tiny portions of the species' former range, where they were subject to the new selection pressures associated with an altered environment. In other words, climate shifts may have been behind the isolation and genetic differentiation of subpopulations derived from an ancestral species that had a much broader distribution. The end result: a swarm of closely related species occupying very small patches of real estate.

Geographic isolation and genetic change may have been augmented by another environmental factor, the diversity of soils of southwestern Australia, which ranges from hard, stony, orange laterites to deep white sands that once covered the ocean floor but became part of the land when sea levels dropped. The various soil types have been shaped by a long period of relative geologic stability in Western Australia. There has been no volcanic activity to speak of for millions of years, no colliding tectonic plates, only modest ups and downs of the coastal plains fringing the outer edge of the continent. As a result, the forces of erosion have worn the land down, leaving behind a mosaic of poor soils with an assortment of chemical deficiencies.

One of the key elements missing from the soils of southwestern Australia is phosphorus, which has been slowly leached out of the ground and carried away to the ocean. Phosphorus is very important to your average plant, which is why the wheat farmers of Western Australia apply phosphorus-rich fertilizers liberally to their fields. Without these amendments, their wheat crops would be even less impressive than they are. Some ecologists believe that good soils (those with plenty of phosphorus and nitrogen) enable a few superspecies of plants to take over entire regions. These species take advantage of the good growing conditions at their disposal to a greater extent than their would-be rivals, blanketing Earth and leaving no room for even slightly less rapid growers. The result: near monocultures of sugar maples or lodgepole pines or, on a smaller scale, cattail reed beds. But when rapid growth is not possible, as is the case for phosphorus-deficient soils, then no one species can hog the space; instead, a diversity of specialists capable of dealing with the particular difficulties of a region can become established. Over time, these plants can adapt to nutrient-depleted soils, and in Western Australia some have done so, making a purse out of a pig's ear, so to speak. Many of these plants not only tolerate remarkably low levels of phosphorus, they actually find higher concentrations downright poisonous. As members of population X have gone about adapting to the specific selection pressures created by the unique pattern of climatic *and* soil conditions in one area, they presumably diverged from other once-related populations. If this scenario were repeated over and over, descendant populations from one ancestral population could have radiated into a battery of ecologically different groups, leading cumulatively to the great numbers of plant species now growing in southwestern Australia.

Still other factors besides climate change and soil diversity may have contributed to the evolution of Western Australia's magnificent plant diversity. For example, the long period of Australian geologic stability may have contributed to the prolonged survival of at least some of the plant species that antedated the relatively recent radiation of the Proteaceae and some other plant groups. Tens of millions of years ago, a very broad band of southern Australia supported a mild temperate rainforest, some of whose constituents still hang on as relicts in modern forests located in the far southwest. The old rainforest genera are typically represented by one or two survivors, unlike the ranks of banksias, dryandras, acacias, and more, most of which have emerged as new species only in the past few million years.

But the currently favored explanation for the large number of plant species in southwestern Australia revolves around its historically variable climate and mosaic of soil types. If this dual-factor hypothesis is correct, then other locations with a history of fluctuating climate and a mix of impoverished soils should also be floral hotspots. Other hyperbotanical areas do exist, which is why one can find 44 percent of the world's vascular (that is, "typical") plants packed into regions that total just 1.4

percent of Earth's terrestrial surface. Among the plant hotspots similar to that of southwestern Australia are a number of temperate, coastal regions with mild Mediterranean climates, including southern California, southern Chile, southwestern South Africa, and the countries surrounding the northern Mediterranean Sea itself.

When one looks at these other hotspots, both similarities and differences with Australia become apparent. On one side of the equation, the Cape Region of South Africa does share a similar climatic history and soil ecology with Australia. The South African climate apparently became much more arid (due to ocean current shifts) over the past 7 million years or so. In addition, Cape soils are nutrient-poor. These features could have led to the creation of the same kind of temporally shifting habitat types that have characterized the history of Western Australia. Moreover, as is true for Australia, the Proteaceae are an important contributor to floral diversity of South Africa. That the two continents have large contingents of this group of plants stems from their shared geographic history, which featured a period when they were united with Antarctica and South America in the ancient supercontinent Gondwana. When Gondwana broke up, various kinds of Proteaceae went their separate ways, with a handful of genera evolving into more than three hundred species in South Africa after a geologically brief but exuberant burst of speciation within the family.

Although the climate shift/poor soils hypothesis for the botanical hotspot in southwestern Australia receives some support from the South African situation, the same cannot be said for California, one of the other Mediterranean floral hotspots. California soils are not particularly poor, nor has the area become more arid in recent geologic time. Therefore, either the climate shift/poor soils hypothesis is flat wrong or it has only limited applicability, with a different set of factors responsible for the botanical riches of California's Mediterranean region. This state, with its vast array of plant species, is far more rugged and mountainous than pancake-flat Australia. The mountains and canyons of California provide a host of geographically isolated peaks and pocket valleys, which may have facilitated the formation of new species there. In addition, the wide altitudinal range offers a host of specialist niches for plant species, niches that are largely absent in Western Australia, which is short on mountains and canyons. Thus, the distinctive topography of California may have something to do with both the production of new plants and their persistence, an argument that could be tested by examining other temperate zone hotspots.

In fact, Mediterranean Turkey and Greece, as well as South Africa and Chile, also are blessed with extensive mountain systems, which could help account for the extreme plant richness in these regions. If so, the explosion of plant species that has occurred in Western Australia, where the hills rarely exceed 500 meters in height, becomes all the more remarkable. (It is true, however, that the one substantial mountain range in the southwest, the Stirling Ranges, is an exceptionally rich botanical locale, suggesting that the availability of mountain peaks, where plant populations can become isolated, promotes speciation everywhere.)

No matter what the evolutionary and ecological reasons for the formation of hotspots, knowing about these places should help those attempting to preserve as much as possible of the world's remaining living things. Rather than focusing on a single species, many conservation groups would like to save examples of entire ecosystems, such as those that are genuine hotspots whose protection would save rich assemblages of organisms. Figuring out which areas will provide the greatest biodiversity bang for the conservation buck is a tough one, particularly because we have to ask whether an area known for, say, the richness of its flowering plants also supports unusual numbers of, say, mosses, liverworts, or fungi. In other words, are some spots valuable across the board or not?

Conservation biologists are hot on this research trail, knowing just how important it will be to get it right, and sooner rather than later. How convenient it would be if one could quickly survey an area with respect to one distinctive and conspicuous group, such as the orchids or the butterflies, and thereby get a good estimate as well of the number of ferns and beetles and birds and tree species in the same region. Although some studies have found that the number of species in one group does predict the number in a different group, in general the results to date suggest that, alas, regions can be superrich in one taxonomic group, such as butterflies, but not unusually diverse in other groups, such as moths. In one such study, a research team divided all of Great Britain into 10 kilometer–by–10 kilometer squares and then added up the number of species of birds, butterflies, dragonflies, liverworts, and aquatic plants found in each square, something that one can do in Great Britain, with its longstanding tradition of tolerance for eccentric naturalists. These enthusiasts have generated species lists of everything from liverworts to butterflies. The cataloguers used the lists to identify the hotspots for a given group, which they defined as those squares in the top 5 percent in terms of species counts. In general, the overlap between any two groups was less than one in four. For example, just 12 percent of the bird hotspot squares were also hotspots for dragonflies.

Other findings also lead us to the conclusion that an area may be rich in one set of organisms but poor in another. Thus, orchid diversity is clearly not necessarily a solid indicator of overall plant diversity, judging from the fact that California has a mere thirty orchid species or thereabouts, despite vast numbers of other plant species. Apparently, we will have to measure biodiversity in area after area, the old-fashioned way, without too many shortcuts. And when all is said and done, some regions will have an abundance of group X and others will be notable for group Y. Then someone will have to sit down and decide whether diversity in group X is somehow more worthy of saving than diversity in group Y. It won't be a happy choice. I know I would take a special interest in the orchids of an area, but my colleague David Pearson would give special weight to the tiger beetles, given his fondness for this group, a quirk that he shares with a surprising number of other beetle enthusiasts.

When Norman Myers and his crew attempted to locate hotspots of biodiversity worldwide they focused on the vascular plants and the verte-brates. It was they who identified twenty-five hotspots covering 1.4 per-cent of the land surface of Earth and containing nearly half of all the vascular plants as well as about a third of all the birds and mammals. For the sake of argument, let's imagine that the powers that be can agree that hotspots with this sort of biodiversity are worth preserving even if other areas may be better for the insects or the nonvascular plants (such as mosses and liverworts). Just what would it cost to protect the twenty-five hotspots? Myers and company estimated an outlay of $500 million over five years, for a $2.5 billion total. This amount is hardly pocket change, but on the other hand, as Myers also has pointed out, governments do lavish cash on certain constituencies, cash that just might be better spent elsewhere. To take one example, Western nations pay their fishing fleets about $20 billion a year, propping them up with major handouts because persistent overfishing has made so many of the world's fisheries uneco-nomic. These subsidies have the most unfortunate effect of encouraging the artificially bloated fishing fleets of the world to continue depleting what little is left of the remaining fish stocks. If one could redirect just a small part of these counterproductive subsidies to land purchases and protection, we could meet the estimated annual cost of Myers's biodiver-sity protection plan *and* help protect the future of the world's fisheries as well. Sounds like a win-win proposition to me.

Or perhaps we could persuade the U.S. Department of Defense to forgo one B-2 bomber a year for five years and invest the savings in the purchase of the top twenty-five biological hotspots of the world. If so, we would have somewhere between $3 billion and $5 billion to work with, an amount far in excess of what Myers and company say is required to protect some of Earth's most wonderfully speciose places. However, you and I know the Defense Department is not going to give up even one ultrabomber, so it's best not to think about it further, lest serious depression set in. Instead, what about setting yourself a challenging orchid-locating goal, say, finding all the nine species of flying duck orchids that are illustrated in Hoffman and Brown's guide to the orchids of Western Australia? This task will keep your mind off defense budgets while also encouraging you to spend hours wandering through nature reserves from Kalbarri National Park well to the north of Perth out to Cape Arid National Park in the east. Your trip will take you from one corner of the twenty-fifth biodiversity hotspot in the world to the other. You can get started in early September and keep going until mid-November. I recommend taking advantage of the biological wonders of the world while they last.

Were B-2 bombers to be phased out in favor of conserving what is left of natural areas in southwestern Australia and other biodiversity hotspots, the result would be improved protection for a great many orchids, among other things. One has to acknowledge, with some pride actually, that the value of such an investment would generate few if any gains in national security, global economics, or medical advances, at least as far the orchids are concerned. This vast group of plants is almost entirely free of usefulness, as measured in human terms.

Yes, the cut-flower trade generates some cash for a few orchid growers. The corsage that Susan Bechtold pinned to her dress forty-five years ago probably contained a pretty *Cymbidium*, *Dendrobium*, or *Cattleya*, to name a few of the genera that are still commercially reared for their large, waxy flowers. Although orchids are considered suitable accessories primarily for women in our society today, it was not always so. The British colonial secretary Joseph Chamberlain (father of the better-known Neville Chamberlain, the author of the policy of Nazi appeasement) is said to have done much to popularize orchids in Britain at the turn of the twentieth century by always appearing in public with a showy orchid in his buttonhole. A natty dresser, the colonial secretary also sported a monocle.

The vanilla orchid also has commercial value as the only orchid to produce something edible for those of us living in Western society. When you and I consume high-quality vanilla ice cream, the little black specks in our dessert are fragments of the tiny seeds of the vanilla orchid, and the aromatic flavor comes from fermented seed capsules of the plant. Although the value of vanilla today is entirely as a flavoring, previously

7

Orchids and

Conservation

people wanted to believe that it could cure a wide range of illnesses as well as possessing aphrodisiacal potency. Perhaps this is why the Aztec chieftain Monteczuma insisted on having access to fifty pitchers of chocolate laced with vanilla each day. One presumes that the pitchers were not large.

The notion that some orchids enhance sexual activity has been around for a long time. The supposed aphrodisiacal effects of the orchid *Dendrobium nobile* resulted in its large-scale cultivation in China during the Han dynasty (207 BC to AD 220). Chinese orchid farmers harvested and processed the rootlike pseudo-bulbs to create a concoction that consumers hoped would be sexually stimulating. It's too bad that those who downed preparations of *Dendrobium* extract were unaware that they were ingesting a solution rich in alkaloids, actually poisonous in large doses.

In a later era and in Europe, the bulbs of certain other orchids were savored for their supposed aphrodisiacal effects or, alternatively, as a suppressant of sexual urges, depending on the psychological needs of the user. Because the bulbs or tubers of some orchids look like mammalian testicles, medieval herbalists assumed that eating these plant parts must have reproductive consequences for the consumer. The very word "orchid" is derived from the Greek word (*orchis*) for testis, which is the root, so to speak, of the unhappy word "orchidectomy." Despite the superficial connection between orchids and testes, no orchid product has ever been rigorously examined for aphrodisiacal properties as far as I know

A few orchids have been said to have medical value. In eastern Europe, even today, the yellow lady's slipper orchid is grown on a modest scale for the roots, which are dried and used to prepare a medicine with "spasmolytic, thymoleptic, and diaphoretic" properties. This herbal remedy supposedly combats insomnia, anxiety, and pain. Consumed in quantity, however, the root causes hallucinations, which would seem to be a good reason to try something else for insomnia, anxiety, and pain.

The fact of the matter is that most orchids are simply not utilitarian but instead provide aesthetic pleasure only, pure and simple. The hobbyists and horticulturalists who grow *Catasetum* or *Cymbidium* or the like do so because they enjoy looking at their plants and relish the challenges associated with keeping them alive and using them to produce attractive hybrids. I recently had the pleasure of attending a meeting of the local orchid club, whose members had lined the tables around the room with their latest triumphs, here a little drooping stalk covered with several dozen tiny greenish blooms, there a stark white flower with one dramatically whiskered petal, and next to it an almost garish creation with immense purple and white flowers. The club president spoke dreamily of an upcoming trip to a major orchid show, where it would be possible to spend a small fortune to acquire fine orchids from all over the world.

The custom of growing tropical orchids at home in the temperate Northern Hemisphere may have begun when a gorgeous Brazilian orchid was used as packing material for other plants exported from Brazil to England in the early 1800s. This specimen triggered a craze for tropical orchid species among upper-class Englishmen. (You should know, however, that many different accounts exist of the early history of orchid cultivation in Europe.) A Sir Joseph Banks pioneered orchid horticulture; his *Vanda roxburghii*, a Southeast Asian species, flowered at his estate in 1817. William Cattley was also in on the action early, getting a specimen of *Cattleya labiata* to bloom in his Suffolk hothouse in 1818. The genus was named for Cattley by the British botanist John Lindley, who no doubt gave his horticultural friend much pleasure by immortalizing him taxonomically.

The beauty and rarity of many tropical species, and the difficulty in growing them, accounted for much of their appeal to status-conscious purchasers. Only relatively wealthy individuals could afford to buy these exotic, imported orchids, let alone keep them alive in a heated greenhouse. In the mid-1800s, the gorgeous Southeast Asian orchid *Vanda coerulea* was sold by Veitch & Sons, a major British nursery, for £300, a non-trivial sum even now and a small fortune then. The money available from trade in wild orchids enabled nurseries to hire collectors who wandered the globe in search of novel species of great commercial value. Veitch & Sons hired people like Thomas Lobb, who found *V. coerulea* and many other beautiful orchids during a four-year expedition through the jungles of Burma, Java, and Sarawak, among other places. Lobb attempted to keep the location of his finds a secret from his fellow collectors by preparing quadruplicate sets of dried specimens of the orchids he collected. He gave each specimen sheet of the same species a different collecting locale either in Java, Borneo, the Malay Peninsula, or Luzon in the Philippines. Needless to say, this commercially motivated practice did not endear him to the botanists who later came to work on the orchids that Lobb brought back from Asia.

Thomas Lobb survived the rigors of jungle collecting, although on what proved to be his last trip he lost a leg "due to exposure," according to Merle Reinikka, whose book on the major figures in orchid biology makes room for a short biography of Lobb. The loss of the leg in the Philippines ended Lobb's collecting adventures and sent him into retirement in Cornwall, where he remained at home except for a single visit to his old employer, James Veitch Jr., with whom he doubtless intended to reminisce about orchids. (According to Luigi Berliocchi, Lobb "was an extreme introvert whose conversation centered exclusively on the orchid and its ways" [*The Orchid in Lore and Legend*, p. 80].) However, during Lobb's stay, the wealthy nurseryman Veitch suddenly died, after which the retired, one-legged orchid collector went back to Cornwall for keeps.

Today, commercially driven orchid collecting from wild populations has declined, although unfortunately it has not ceased entirely. For example, when a nursery owner from Virginia paid a rural Peruvian family $6.50 for a specimen of what proved to be a previously unknown (and very striking) species of *Phragmipedium*, the locals promptly went to work collecting the slipper orchid with a vengeance. Their skill in removing the plants became evident when the nursery owner returned to secure additional plants for research and horticultural breeding. He found the original orchid site stripped, not a single specimen remaining. Whether some populations of this orchid have survived the initial rush to cash in on its novelty and evident rarity is a matter of debate (see the forum on *Phragmipedium kovachii* at theorchidsource.com). Let us be hopeful.

Happily, many enthusiasts recognize the need to conserve wild orchids in their natural state. Rather than ostentatiously displaying orchids ripped from a rainforest, many persons now own specimens that began life as seeds taken from a greenhouse plant and induced to grow into domesticated pot-dwellers, thanks to horticultural tricks of the trade that make it possible to rear some orchids on a grand scale. These tricks are necessary because growing an orchid from scratch is not easy. Orchid seeds are among the smallest of plant seeds. Their tiny size and corresponding light weight facilitate wind-aided dispersal, but this benefit comes at a cost. The seeds are so minute that they have almost no nutrients to support the growth of an embryo. Because they cannot germinate and grow on their own, they rely on certain fungi, which come to the rescue by insinuating their thin filamentous hyphae into the seeds. Once inserted, the fungus supplies the seed with water and nutrients, which are funneled in via the tube-like hyphal thread. Eventually, portions of the fungal filament begin to bloat and break apart. As they do, orchid mitochondria assemble at the site, where these microfactories go to work converting fungal nutrients into energy that orchid cells can use to survive and multiply.

At least some of the fungi involved customarily make their home in decaying wood, which they colonize and decompose. Thus, they are not restricted to orchids, although they are usually essential for orchid seed germination. As development of the embryo proceeds, mycorrhizal fungi also continue to contribute to the growth of the orchid, which forms a mass of undifferentiated tissue called a protocorm. If all goes well, this stage eventually becomes large enough to sprout roots and send out self-sustaining, photosynthesizing leaves (in those orchids that produce chlorophyll-rich leaves). If one "feeds" radioactively labeled carbon to mats of fungal hyphae in the soil outside a growing orchid, the radioactive carbon begins to show up in the carbon-containing constituents of the orchid's cell walls. In contrast, if one permits the leaves of an orchid to pick up labeled carbon, the radioactive element does *not* move within the plant to the fungal hyphae and out to portions of the fungus growing in the soil. Thus, it appears that the fungus gives but does not receive, so that the orchid derives all the benefits from their special relationship. I wonder if it will eventually be shown that the orchid "tricks" the fungus into inserting its growing filaments into its seeds and roots, perhaps by mimicking the chemical cues provided by the normal substrates beneficial to the fungus, such as decaying wood.

In any event, the point is that most or all orchids are dependent in one way or another on the seed- and root-infecting fungi that choose to take up residence inside their cells. Before the vital connection between particular fungi and orchid seed germination was known, horticulturalists wishing to grow orchids from seed, rather than extract them as mature plants from Javan jungles, had to deposit seeds on the roots of an established plant of the same or similar species. In this way, some seeds acquired the right fungal partner and so could germinate, but the process was very much a hit-or-miss proposition, resulting in a low success rate. After more was known about orchid seed germination in nature, some horticulturalists experimented by putting seeds in sterile glass containers along with fluids containing a mix of salts (fertilizers) and sugars normally supplied by symbiotic fungi. In 1919, Lewis Knudson, a botanist at Cornell University, succeeded in producing seedlings of a *Cattleya* orchid after germinating the seeds in artificial media and growing the resultant protocorms. The procedures that Knudson used have been modified and widely disseminated, enabling ordinary orchid enthusiasts to grow their favorites at home from seed, which has surely helped preserve species that would otherwise be at risk from unethical field collectors. When the homegrown specimens flower in captivity, they can be used to produce hybrids by placing pollen from one species on the pollen-receiving stigma of another. Because manufacturing hybrids via hand pollination is relatively easy, many novel orchids have been produced, some of which have flowers that appeal to orchid fanciers and the judges of orchid shows.

Thanks to the modern techniques of orchid propagation, most orchid growers are happy to support orchid conservation in the wild. Alas, their cooperation is not enough to solve all the problems that confront this family of plants. Take, for example, the elegant spider orchid (*Caladenia elegans*). This extremely handsome plant is not threatened by avaricious collectors determined to take possession of it. And yet, *C. elegans* is unquestionably rare, with a current population of fewer than five thousand individuals scattered about in a handful of rural locations several hundred kilometers to the north of Perth. The plant is so scarce that it has been selected for a recovery plan drawn up by the orchid specialist Andrew Brown and other plant conservationists who work at Western Australia's Department of Conservation and Land Management (CALM). Reading the plan is sobering because it is quickly apparent that the elegant spider orchid's rescue will require coordinated action on more than one front.

The problem at the top of the list is habitat loss, which comes as no particular surprise. Environmental destruction is the major reason so many species, plant and animal, are in trouble right round the world.

It is instructive that the lovely Brazilian orchid, *Cattleya labiata*, whose importation to England in the early 1800s had much to do with the craze for tropical orchids, had already largely disappeared in the wild by 1837. The cause was not collecting but wood cutting and urban sprawl in the vicinity of Rio de Janeiro. A Dr. Gardner wrote, "The progress of cultivation is proceeding so rapidly for twenty miles around Rio, that many of the species which still exist will, in the course of a few years, be completely annihilated, and the botanists of future years who visit the country will look in vain for the plants collected by the predecessors" (quoted in Reinikka, *The History of the Orchid*, p. 25).

Thus, when Andrew Brown began his work with the elegant spider orchid, his top priority was to see what was left in terms of suitable habitat for the species. As he surveyed remnant patches of native vegetation surrounded by wheat fields and sheep pastures, he was thrilled to discover one site with several thousand members of this species. Less than two years later, the area had been illegally cleared, completely wiping out that large and healthy population. The farmer responsible was charged, tried, and convicted, but he was in effect simply told not to do it again, given a mere slap on the wrist despite the fact that the law provides for much stiffer penalties in such cases. What was once prime elegant spider orchid habitat is now just another dusty Australian pasture with essentially no chance of supporting any elegant spider orchids ever again.

Elsewhere, another population of this lovely yellow orchid is being gradually nibbled to death by road graders as they maintain a dirt road that impinges on orchid habitat. Road graders are a common sight in Western Australia as they rumble down the unpaved roads that crisscross rural landscapes there. The operators of these machines do an excellent job, so much so that one can usually fly safely down an Australian dirt road connecting one hamlet to another at 50 miles per hour. But grader operators often seek not merely to smooth but to widen the road under their control, scraping away just a wee bit more roadside vegetation each time they go to work. As a result, roads that might see a few dozen cars and trucks each day are often as wide as a four-lane highway, more than enough to permit two grain trucks to pass one another at top speed. The extra width comes at the expense of native plants, which take advantage of the attractive Australian custom of preserving uncleared roadside strips that separate the roadway from nearby agricultural lands. These long thin verges are supposed to be protected reserves, and rightly so, because they often provide the only sanctuaries for native plants for miles around, spidery avenues of green in an ocean of yellow wheat fields and gray paddocks. When road reserves are reduced in width, they lose some of their elegant spider orchids and other dinkum (true) Australian plants.

The fact that native habitats have been reduced to mere roadside verges in so much of southwestern Australia is largely a result of introduced mammals. When one thinks of mammalian imports to Australia, European rabbits, European red foxes, and European cats immediately come to mind, but I will argue shortly that these are not the worst offenders. First, let me assure you that I am no apologist for Australia's imported rabbits, foxes, and cats. Cats were brought over with the initial wave of convict colonists in the late eighteenth century, and rabbits and foxes started to plague the country in the mid-nineteenth century when released to provide sport for Australian gentlemen. The negative press that this trio eventually received is entirely deserved, as rabbits surely compete for food with the far more delightful Australian herbivores and the red fox and cat make meals of the natives. In fact, red foxes and cats are such highly efficient predators that some persons have proposed that these introduced killers are behind the extinction of some or all of the sixteen Australian mammals that have disappeared in the past couple of hundred years. These sixteen species constitute about half of all Earth's mammals that are known to have gone under during this period.

The fox-cat hypothesis for Aussie extinctions has been tested. If it is true that European foxes and cats are lethal enemies of native mammals, then some of the currently rare and endangered species should do better if there were a way to rid an area of the introduced predators. Happily, a combination of lethal baiting, shooting, and trapping has freed some reserves in Western Australia from foxes and cats. Foxes are especially vulnerable to eggs and dried meat baits injected with the lethal 1080 poi-

7.1

An escaped agricultural lupine lurking behind a white spider orchid in what was once rich orchid habitat near Lake Yarra, Western Australia.

son. In one study, all forty-five radio-collared foxes in a baited area were dead less than two months after the first baits were put out.

Cats offer a tougher challenge because they rarely feed on carrion and in the past had to be tracked down one by one. (Recently, CALM has come up with a bait irresistible to these dreadful creatures, for which we can be thankful.) During previous control efforts that focused on getting rid of foxes, feral cats increased in number because they were no longer at risk of becoming a menu item for a fox. An increased number of cats led to increased predation on the small to midrange native mammals. However, when introduced foxes *and* cats have both been sent to their reward, the bettongs, numbats, and other strange, wonderful, and endangered Aussie mammals can be safely reintroduced into nature reserves, where they will flourish—so long as their unnatural predators can be kept from returning. In one project of this sort, the burrowing bettong, which had been utterly exterminated on mainland Australia, was successfully reintroduced to a mainland peninsula that had been fenced off and swept clean of foxes and cats. In seven years, the forty-two bettongs translocated from a predator-free island produced enough surviving offspring to boost their numbers more than sixfold.

But, as I say, although rabbits, foxes, and cats have done dreadful things to Australia's native animals and plants, another group of introduced mammals has actually damaged the environment far more. I point an accusatory finger at the sheep, cattle, horses, and goats of Australia, which started munching their way across the continent soon after the beginning of European colonization. They have altered the country massively in ways that are highly significant for the native vegetation, including the orchid component. Millions upon millions of sheep have grazed all of arable Australia, including most of the southwest, at one time or another. After they had eaten everything edible in their path, some herds were removed, and what were once pastures were turned into wheat fields.

But to this day, sheep are a mainstay of the Australian economy. Whether present or absent, the effects of sheep and other livestock are remarkably long-lasting. The durable nature of the changes these animals have caused on Australian ecosystems stems from the ability of livestock to prepare the way for introduced European weeds, many of which find the disturbed soils of grazing lands an ideal environment for their establishment.

Weeds are, according to the dictionary, "useless, troublesome, or noxious plants, especially those that grow profusely." The primary offenders, it turns out, are plant species that have been intentionally or accidentally introduced from one continent to another. Even agriculturally valuable plants can become weeds when they grow where they are not wanted. So, for example, lupines are a farm crop when they stay within the borders of a farmer's fields, but when these plants, which are widely grown in Australia for their oily seeds, move into nature reserves they qualify as weeds. It is a sad thing to see lupines taking over the open woodland near Lake Yarra, where the escaped crop now towers over the little daisies and white spider orchids that once had the place all to themselves.

7.2

The accidentally introduced South African orchid, *Disa bracteata*, is common throughout southwestern Australia.

Although some Australian plants have become exotic weeds elsewhere, as in the eucalyptus that have escaped into South African heathland and the melaleucas that have become established in Florida's everglades, Australia has suffered substantially from its unwelcome invaders. About three thousand of the continent's twenty-five thousand plants have come from somewhere else. One is that South African orchid, *Disa bracteata*, we found on Mt. Ridley. Seeds of the orchid supposedly came to Western Australia in a shipment of South African goods unloaded at Albany on the south coast. This self-pollinating orchid can now be found throughout much of the southwestern corner of the state, flourishing in pastures but also present in a host of disturbed natural habitats as well. A chunky greenish plant with a multitude of minute reddish brown flowers, this import doesn't set anyone's heart a-racing but is instead just another reminder of how we humans have jumbled plant geography all over the globe.

As weeds far more damaging than the introduced orchid have taken over Australia, they have made it difficult for abandoned pastures to return to their original condition. Worse, pasture weeds are always pushing their way into the remnant patches of natural vegetation. Thus, livestock and weeds have worked hand in hand to fragment the Australian landscape, always tightening the noose around any natural habitats that manage to avoid the plow or bulldozer. Although exotic plants kill less dramatically than a flock of hungry sheep, they can engineer a very convincing death for the native plants with which they compete.

It is ironic that many troublesome foreign plants have been openly welcomed into Australia because Australians, like their counterparts in other parts of the world, have wanted to make their gardens more attractive, their agricultural crops more diverse, and their pastures more productive. One such problem plant is the appropriately named Paterson's curse, an attractive purple ornamental from the Mediterranean region that first adorned Australian gardens in the 1840s but subsequently slipped away from cultivated flower borders to take up residence in 33 million hectares of rangeland, where it pushes out introduced exotic grasses that cows and sheep find far more edible.

The damaging economic effects of assorted weedy imports has been documented by William Lonsdale, an entomologist employed by CSIRO who conducted a general survey of the plants purposefully introduced into the pastures of northern Australia in the mid-twentieth century. Of the 463 grasses and legumes put out in the hopes of boosting grazing productivity, a mere twenty-one have panned out from the perspective of the livestock industry. Nearly three times as many imports (sixty species) went on to cause serious environmental problems. Among the bad guys were seventeen of the twenty-one "useful" introductions, which were ill-mannered enough to grow vigorously where they were not wanted. Indeed, as Lonsdale points out, any plant that can take root well enough in foreign soil to generate an economic benefit of some sort has an equally high potential to do damage by surviving to flourish where it is not "supposed" to grow. Lonsdale sensibly suggests that it might be a good idea to consider the downside of a plant introduction *before* setting a nonnative species free in a foreign land. According to a university report, weeds set Australian farmers back something like A$3.9 billion annually in terms of lost productivity, control costs, and the like. Given the expenses of combating a weedy intruder that has run amok, the economic benefits of having a new plant in places where it is desired had better be great indeed if a net financial gain is to result from its introduction. Although Lonsdale is careful not to be too judgmental, the facts he has assembled speak for themselves: nonnative plants brought in to improve pastures have done far more damage than good in northern Australia.

7.3

The purple Paterson's curse, a plant intentionally brought to Australia, now is a destructive weed that takes over entire pastures.

The same can be said for other parts of the world. The planetary scope of the weed problem is staggering. In temperate North America, for example, somewhere between 20 and 30 percent of our plants are aliens, introduced from Europe or Asia or Australia. More than a thousand non-natives have been recorded in California alone, a state that now hosts any number of Australian eucalypts. A considerable number of these foreigners have proven to be a disaster in various parts of their expanded range. One of the classic examples is the Frankenstein vine kudzu, which was brought to the United States in 1876 and later used on a large scale to combat soil erosion. The vine does indeed cover bare ground quickly, growing up to a foot each day during the spring and summer, sending down new roots at intervals, stabilizing the soil in the process. But it also swarms over everything in its path, whether this be an eroded gully or a shrubby hillside or a young pine plantation. Like a South African vine, bridal wreath, which has sadly found a comfortable new home in southwestern Australia, kudzu kills by smothering, blocking out the sunlight that the plants beneath it need to photosynthesize and thus survive and reproduce. The plant now occurs from Texas to Florida north to Illinois and Pennsylvania, having commandeered some 7 million acres of land. It is no longer welcome in the United States.

Nonnative grasses have been particular import favorites in Australia, the United States, and almost everywhere else, with results that are often ugly. For example, in my homeland, the arid southwestern United States, red brome grass (from the Mediterranean region) and buffel grass (from Africa and India) have been broadcast to provide food for rangeland cattle. These grasses have coped rather too well with the novel conditions they have encountered in their new homes. Invading foreign grasses can displace native species in a very straightforward manner. The newcomers simply produce more seeds than their competitors, which has led environmental restoration experts to broadcast extra seeds of native grasses in areas where exotics have established a beachhead. The countermeasure looks as if it works, but it is of course expensive and requires the production and collection of massive amounts of native seeds.

Here in Arizona, one problem with invading grasses is their tolerance of drought. Even in low rainfall springs, the exotic species grow relatively luxuriantly (by desert standards), then die and dry out, forming a fuel base that sometimes burns during the summer. Wildfires are an evolutionary novelty in much of the desert southwest; many native cacti and shrubs have no defense against incineration and therefore expire when burnt, rather than regenerate in the manner of fire-adapted plants that have evolved in environments where fire was and is a regular phenomenon. Because red brome and buffel grass are fire-adapted, once a wildfire chars an area, the seeds of these introduced grasses survive nicely in the soil. Then, in the next growing season, buried brome and buffle grass seeds germinate freely, forming even denser stands of grass that fuel even hotter fires. In short order, desert cacti and shrubs give way to simple, unnatural grasslands that are far less diverse and interesting than the communities that preceded them.

So what we have right around the world is an unhappy synergism between introduced livestock and introduced weeds. Cattle and sheep disturb the land, whether in Australia, the United States, or Brazil, creating good habitat for weeds, some of which make their way to pastures by accident, while many others are intentionally broadcast by livestock owners. The new plants then are free to make their way off the plantation, so to speak, and some turn out to be good at expanding into natural areas, where they smother native flora, stealing the sunlight, space, or nutrients needed by these far more interesting plants. In southwestern Australia, pasture grasses regularly crowd into roadside verges, and when they do, the native orchids are among the first to disappear.

Whereas weeds tend do their damage by outcompeting native plants for key resources, some introduced fungi can prey directly on their floral victims. Notable among these enemies of Australian native plants is the dieback fungus *Phytophthora cinnamomi*. The genus *Phytophthora* is a large one whose members have a history of making life miserable for certain plants after people have inadvertently carried the fungus into a new land. The most famous example involves a strain of the potato blight, *Phytophthora infestans*, that entered Ireland in the mid-1840s, where it contributed mightily to the terrible famines that killed more than a million Irish by 1850 and prompted the emigration of millions more. To this day, several *Phytophthora* fungi continue to threaten important agricultural crops and valuable trees around the world. The threats grow worse by the year. Recently, a new pathogenic *Phytophthora* was found destroying the roots of the Port Orford cedar, an economically important conifer that grows in the northwestern United States. Another fungus in the same genus causes sudden oak death syndrome, a disease that has swept through the oaks of California and Oregon, killing thousands of trees. As if the destruction of oaks were not bad enough, the fungal microbe also apparently infects many other important forest trees, including Douglas fir and possibly even coastal redwoods, a most unhappy possibility indeed. Just where the new *Phytophthora* has come from remains unknown, although surely it is a nonnative, accidentally introduced to our country.

The *Phytophthora* that has caused no end of sadness in southwestern Australia also is a foreigner, although different authors identify different countries as the source of the fungus. I have read that *Phytophthora cinnamomi* may have been brought in as early as the 1820s by early colonists

7.4

Many species of *Banksia* are highly susceptible to the lethal exotic fungus *Phytophthora cinnamomi*. The plant in the top photo is a healthy *Banksia speciosa*; the plant in the bottom photo has just died, almost certainly as the result of root fungal infection.

who carried cultivated European plants around in soil-filled containers. Or was the fungus imported from Indonesia in the 1930s in horticultural shipments? Or did dieback arrive in the 1950s in the soil adhering to apple root stock? No matter. Once in Australia, the fungus found a whole catalogue of novel victims, among them some eucalyptus trees important to the local timber industry as well as many aesthetic flowering plants. Having evolved apart from the *Phytophthora* in question, Australian plants generally lack capable defenses against the invader. Members of the family Proteaceae are particularly vulnerable to dieback, a disease that occurs when the root-infecting fungus begins to destroy the plant's root cells and thus its capacity to acquire water from the soil. Sometimes, only the outer branches of infected trees and shrubs die back, but in severe cases the strangled plant simply succumbs, collapsing into a pile of cracked gray wood on the ground.

The losses among the Proteaceae caused by the dieback fungus are particularly troubling because, as mentioned previously, southwestern Australia is home to many conspicuous and highly attractive members of this family that are found nowhere else in the world. The banksias, dryandras, hakeas, grevilleas, and isopogons add great color to the heathlands and open forests of the southwest during their flowering seasons. Their destruction diminishes the beauty of these places in ways obvious even to the most casual observer.

Although dieback *Phytophthora* does not usually infect and kill orchids directly, it can influence some species' chances of survival by killing the trees and shrubs that provide critical shade and cover for understory plants. The bottom line is that intact native habitats are essential for orchids. Bulldoze the native bushland, or let exotic weeds take over, or infect the larger plants with dieback fungi—take your pick. Any of the above can be a death sentence for the orchids that once brought their own very special allure to a once intact natural area. Nor have I yet mentioned another kind of man-driven habitat change that has grave consequences for some orchids and other Australian native plants. When the farmers of Western Australia succeeded in cutting down and burning most of the woodlands that once covered what were to become their farm fields and pastures, they inadvertently also changed the water table underneath their lands. The eucalyptus trees that were reduced to charcoal had been performing a vital hydrologic service, which was to draw water from the soil and pass the moisture out through their leaves and into the air. As a result of this process, the water level in the soil was kept low, holding the salts down deep in the earth. When the land was cleared, however, billions upon billions of water-losing leaves were destroyed, the water level rose, and the lethal salts were drawn up into the root zone. Many a once productive field and pasture in Western Australia has been reduced to an oozing, barren, salt-encrusted wasteland capable of growing next to nothing. Long before the situation becomes desperate, agricultural crop production falls, as does the survival of plants in adjacent natural areas. One of the most endangered of all of Western Australia's little orchids is the hinged dragon orchid, *Caladenia drakeiodes*, which has suffered greatly from clearing and increased salinity in woodland patches surrounded by treeless wheat fields and salty pastures.

Underlying almost every case of habitat destruction or degradation, whether caused by bulldozers or invading weeds, is the one root cause of them all: too many people. The human population is immensely demanding of space and resources. Just how demanding becomes clearer when you examine a map of the globe produced by a group of researchers who wished to show the size of the "human footprint" on the land. The map reveals that 83 percent of the land surface must deal with one or more human "impacts," including a population density greater than one person per hectare, agricultural land use, a collection of dwellings, and enough light at night to be detected by satellite sensor. Moreover, nearly 98 percent of all land suitable for rice, wheat, or corn is subject to the human footprint. Those of us living in North America, Europe, and Australia are all "developers" in the sense of being directly employed by the farming, mining, logging, ranching, real estate, and construction industries or else totally dependent on the activities of farmers, miners, home builders, ranchers, and the like. The more of us, the harder we are on our planet.

This axiom can be put to the test in Australia because there are "only" about 20 million or so Australians in a country the size of the continental United States, which accommodates nearly 280 million people. But even though the Australian population is less than a tenth that of the United States, every last square inch of arable land is under cultivation in Australia, as are a good many square inches of borderline semidesert. As a result, almost every year somewhere in drought-prone Down Under, some farmers or ranchers are confronted with the catastrophic loss of their crops or their livestock, leading to appeals for governmental assistance to keep these unlucky, some would say unwise, agricultural producers afloat until it rains again. No matter how marginal the land for wheat or sheep, if it can be farmed at least occasionally, then the land has been developed, generally long ago.

7.5

Salt damage in Western Australia. The rising water table has pushed lethal salty water into the root zone of the eucalypts that once grew along this little watercourse running through a pasture.

In his book *The Red Centre*, the mammalogist H. H. Finlayson noted just how quickly the entire continent was taken over by what North Americans would call ranchers, with their "cattle, horses, donkeys, and camels, . . . sheep and goats and dogs; the great hosts of the uninvited also—the rabbits, the foxes, and the feral cats. The results of all this are hailed by the statistician and economist as progress and a net increase in the wealth of the country, but if the devastation which is worked to the flora and fauna could be assessed in terms of the value which future generations will put upon them, it might be found that our wool-clips and beef and timber trades have been dearly bought" (p. 2).

Finlayson was writing in the 1930s, when less than 7 million persons lived in all of Australia, a figure less than the current population of Hong Kong. Although by the 1930s the number of Australians had grown greatly from the few hundred thousand colonists (plus an undetermined

number of aboriginals) present at the time of Darwin's visit a century previously, the total was still very modest by twenty-first-century standards. Nevertheless, the relatively few Australians of the time had already seen most of their country dramatically changed for the worse. How had this happened? As Finlayson points out, Australia has "no considerable mountain masses with climates sufficiently rigorous to discourage a stockman; few deserts or bad lands supporting any considerable fauna, which will not also support sheep and cattle; no . . . effective barriers to the introduction of domestic animals" (p. 3). Finlayson realized how much livestock amplify the environmental damage that accompanies the human population wherever it goes.

Although biologists knew even by the 1930s that Australia was overpopulated in the sense that the whole country had already been heavily exploited, this view still lacks general acceptance, despite the near tripling of population that has occurred since the 1930s. Indeed, many influential Australians are convinced that their nation is actually *under*populated. For example, former prime minister Malcolm Fraser recently stated that Australia "should be aiming for a population of 40 to 50 million people." Fraser asserted that "to keep the wealth of Australia to about 20 million people alone is probably very selfish" (quoted from the *Sydney Morning Herald*, June 25, 2001). Of course, if Fraser and others wish to practice national charity, Australians could simply donate a portion of the country's wealth to others while keeping their own population constant. On the other hand, if the plan is to have sufficient manpower to repel Indonesian or Malaysian hordes at some point in the future, then even 50 million citizens probably won't do the trick. Perhaps this is why certain economists and senior fellows at Australian think tanks have argued that Australia needs a population of several hundred million to hold its own against the far more populous nations to the north.

In *The Future Eaters*, a fine book on the ecological history of Australia, the biologist Tim Flannery makes a strong case that Australia needs to lose population, not gain yet more millions. He points out that 70 percent of Australia's cultivated lands have lost productivity in recent years, a fact that ought to be wake-up call for the "populate or perish" boosters. The simple reality is that, as Finlayson knew long ago, there are no ungrazed, unfarmed, and undeveloped regions of Australia worth bringing into production today. A smaller Australian population of, say, 6 to 12 million, Flannery's target, would put significantly less stress on the limited arable fraction of the continent, so that it would be far easier for individuals in the future to experience a standard of living commensurate with the current good life that many Australians now enjoy.

Regrettably, Australian politicians are no different from the vast majority of politicians elsewhere in steadfastly ignoring the need for social policies that would encourage population stabilization, let alone a reduction in their country's numbers. So the root cause of Australia's conservation problems will not be addressed at a governmental level anytime soon. Meanwhile, what are some of those approximately 20 million Aussies doing to promote the preservation of their biological heritage? Here at last we can discuss something positive, because many Australians are well aware of the problems their homeland faces. This awareness has translated into any number of useful programs, some of them involving government agencies and employees. For example, whereas land clearing was actively encouraged and vigorously subsidized throughout Australia for many years, such is no longer the case. Australians have to some extent overcome the pioneer ethic of settling and clearing the land, a homesteading tradition with which some living Aussies have had personal experience. I once met an elderly gentlemen at the Stirling Range Caravan Park who told me about his youth spent grubbing out poison pea shrubs on his father's homestead. The years had not dimmed the memory of what hard work that had been, all necessary to make the land safe for grazing sheep, which would have been killed had they nibbled a leaf or two of the lethal poison pea. After the war, land clearing by pick and axe gave way to the mechanized form of landscape reorganization. And bulldozers continue to scrape their way across Australia, but the number at work has declined sharply over the past decade, in part because of new legal restrictions designed to preserve the remaining fragments of uncleared land.

In addition to governmental regulations designed (in theory) to penalize persons for unauthorized land clearing, some new programs are in place that try to encourage responsible land ownership. The Land for Wildlife program makes it possible for landowners to voluntarily register portions of their private land with an agency whose personnel can help them manage the real estate in question for conservation purposes. Properties that are selected for the program can, for example, receive financial assistance for fencing off sensitive areas, such as stream banks or remnant patches of woodland, and they can get advice on how to restore degraded habitats for the benefit of native plants and animals. Given that species at risk often occur on private property, a program that encourages landowners to protect their land makes sense. As of late 2001, a total of 100,000 hectares owned by six hundred Western Australians had been registered in the program.

Opportunities also exist for revegetating degraded patches and for mass plantings of native trees and shrubs of possible commercial importance. In 2002, wheat belt farmers were offered at no cost a total of about 2 million melaleuca seedlings, which may produce oils of value while also helping to restore lands that have been badly harmed by previous farming schemes. Admittedly, participants in various planting experiments and Land for Wildlife are free to leave the programs whenever they so choose. If it were otherwise, few farmers and graziers would sign up.

Moreover, when owners sell their land, the new property holder need not honor the previous arrangements and is free to bulldoze the remaining native patches the day after securing the deed. For truly long-term conservation, Australia needs public reserves owned and permanently managed by competent conservation agencies.

To this end, starting in 1998, Western Australia's CALM has invested substantial sums toward land acquisition, most notably in the vast desert chaparral well to the north and east of Perth, the so-called Gascoyne-Murchison Rangeland. This region covers an area of about 34 million hectares (roughly 85 million acres), or, looking at it another way, a parcel of land somewhat larger than New Mexico, the fifth largest state in the United States. In 1998 the rangeland already contained a number of large reserves, but these primarily featured rocky regions and spinifex sandplains, in other words, places where sheep and cattle grazing are impossible, another example of the willingness of people to put "worthless" land into reserves. The result, however, was that many special habitats within the Gascoyne-Murchison Rangeland were not represented in the then current reserve system. Instead, the biological wealth was "owned" by the very few persons and companies in charge of about 250 huge grazing leases. The good guys at CALM had managed as of early 2002 to buy out several lease holders and parts of some other properties as well, acquiring 3.7 million hectares of diverse habitats in the process, an area roughly equivalent to two New Jerseys. Sheep no longer need apply. (In contrast, when our government created a number of new national monuments at the start of the twenty-first century, it did so with the proviso that these areas would remain available for business as usual, including heavy grazing, despite their great biological and archaeological value.)

One of the Western Australian properties now under protection is Muggon Station. When my wife and I passed the turnoff to the station while traveling one of the back country tracks through the Murchison district, we marveled at the remoteness of the homestead and the sense of isolation that the station residents must experience. Although the countryside was beautiful, the barren pebbled desert plains and rows of acacias, many dead, some alive, did not seem to constitute the best possible sheep-grazing habitat. Indeed, less than 10 inches of rain fall in an average year on this desert land. Perhaps this was why CALM was able to acquire the Muggon lease for less than a half million Australian dollars (a figure that at the time was equivalent to about US$250,000). The Muggon lease covers 450,000 hectares, typical for a sheep-grazing property in this part of the world. The purchase price comes out to about US$0.25 per acre, so that CALM has been able to protect a great deal of land for very little.

The Western Australian rangeland that has been retired from commercial grazing will, however, need costly management in perpetuity to deal with the feral goats, camels, donkeys, foxes, cats, and other exotics that now live in the arid center of Australia. Even so, I let myself feel a little twinge of pleasure at the thought that the native plants in this vast and biologically rich desert will now have at least a fighting chance. Tony Brandis, CALM's rangeland coordinator, tells me that within two years of removing livestock from the land, Muggon's plants had rebounded beautifully. The area has apparently become the envy of the neighboring pastoralists (some of whom may admire vegetation largely in accordance with its value as sheep fodder). Among the beneficiaries at Muggon and elsewhere are the native grasses, bindiis, bluebushes, poverty bushes, and mulla mullas, but not orchids, which do not constitute an important part of the flora in the highly arid parts of Western Australia. Orchid conservation will have to be promoted by land protection in the wetter, agriculturally richer regions in the southwestern portion of the state.

As I mentioned earlier, Andrew Brown's team at CALM has been hard at work in the southwest, trying to determine which orchid species are at risk. At the moment, ten species have been designated as vulnerable, thirteen as endangered, and another ten are on the critically endangered list. Having established which orchids need help, the next step is to save or even improve the most important patches of native habitat for the threatened species. Because farmland in the southwest has much more commercial value that rangeland in the arid Murchison, outright purchases of orchid-rich land are generally prohibitively expensive. Thus, Brown and his coworkers have devised some cheaper ways to preserve the habitat critical for a given orchid's conservation. For the feral pigs that plough up the soil in plots where the elegant spider orchid grows: lethal baits containing the poison 1080. For the livestock that graze and trample elegant spider orchids on a rancher's private property: an exclusionary fence. For the road graders that slice away at the roadside verges where the orchid has a foothold: a little get-together with the Shire Council to emphasize the need for grader restraint. For the gravel pits and other heavily degraded sites from which the orchids have been utterly eliminated: a dose of restoration ecology with the planting of a diverse set of native shrubs to re-create an environment congenial for the orchid *and* its pollinators. In some places, orchids have recolonized restored sites on their own, adding new individuals to populations that need every possible recruit.

One way to restore orchids to areas from which they have been lost might be to take seed collected from existing populations and then hold it for dissemination in suitable habitats. Research conservationists have established that drying and then freezing orchid seed in liquid nitrogen works well as a means of storing seed for later use. Orchid seeds that have been treated in this manner sometimes actually germinate better than fresh seed taken directly from a living specimen. But, and this is a big qualifier, this rule applies only to seed that has access to the mycorrhizal fungus necessary for germination, which requires that the fungus be frozen with the seed (possible in some cases), or that fungusless seed be broadcast in areas where the critical fungus occurs naturally, or that the seed be infected with a fungus that has been grown separately. This last possibility has been explored in the United States by Scott Stewart and Lawrence Zettler, who succeeded in growing seedlings of a Floridian swamp-loving orchid, *Habenaria repens*, in the lab from seed placed on agar plates that had been inoculated with a specific fungus. They are hopeful that the technique may be used to reintroduce the orchid to wetlands being returned to something resembling their previous natural state.

Because restoration of environments with their full complement of flora, including the orchids, is fraught with pitfalls and disappointments, it is best to try to keep orchid-rich natural habitats intact and free from degradation. But it ain't easy. As noted earlier, environmental integrity in southwestern Australia (and elsewhere) comes under assault from many forces, especially invasive weeds. Controlling these plants is something of a nightmare given their resilience and high reproductive output. Broadband herbicides can be used in the right circumstances, provided that native plants are not incidentally sprayed. Digging out damaging weeds by hand can also work, although a certain degree of masochism is required given that the worrisome invaders tend to be extremely prolific. Moreover, many weed seeds do well in disturbed soil, so that yanking out adult plants and turning the soil over in the process may merely activate the production of the next generation from the earth's seed bank.

Not the least of the problems associated with attacking weeds lies in the financial cost of trying to do it right, as can be illustrated by one restoration project in Kings Park in Perth. Here large areas of more or less native vegetation coexist with moderately degraded sites, totally exotic lawns and fountains, and formal botanical gardens. The park managers are trying to remove nonnatives from some of the more damaged natural areas in order to replace them with appropriate native species. The job ahead is intimidating; of the 465 plant species recorded at the park, 149 are exotic aliens. To remove the introduced pine trees on steep slopes, huge cranes have to be maneuvered into position. After the big pines have been hauled out, workers must go after unwanted smaller plants by hand. The largely barren soil that remains has to be covered with jute or coconut fiber matting to prevent erosion. Then, just before replanting with native seedlings, the matted slopes have to be sprayed with the herbicide Roundup (also known as glyphosate) to prevent weeds from reestablishing themselves. Slits are cut in the mats and native plants inserted through the slits into the soil beneath. The cost of repairing a mere 50 square meters is between A$160 and A$185 for the matting plus about A$90 worth of steel pins or starch pegs to hold the cover in place. Then there is the expense of paying those who do the hard work of laying the cover and replanting the slopes (four man-hours of labor are required for each 50 square meters). Fixing what needs to be fixed does not come cheap or easy.

Despite the intimidating and expensive obstacles to effective weed control, some brave souls have entered into hand-to-hand combat with damaging exotics in various parts of the world. In a Hawaiian reserve for the endangered Hawaiian stilt, a shorebird, the U.S. Marine Corps has physically removed introduced red mangrove from 20 acres of mudflats. The mangrove, a kind of aquatic kudzu, was originally brought to Hawaii for soil erosion control in coastal areas destroyed by livestock and humans. The repair job took twenty years, thousands of hours of volunteer labor, $2.5 million of contract labor, and much heavy equipment. The restoration project worked, but keeping red mangrove from reinvading the cleared reserve will require eternal vigilance.

In Western Australia, Peter Day, a one-man Marine Corps unit, has also devoted twenty years at approximately three hundred hours per annum in his attempt to eliminate *Watsonia* lilies from his community. Lilies in this African genus have been domesticated by horticulturalists and distributed to gardens around the world. Many of the fifty-two species that occur in South Africa are big handsome plants somewhat reminiscent of gladioli, one of which is mimicked by a deceptive South African orchid (see Chapter 4). Unfortunately, some of these exotics were not content with life as a good-looking ornamental in urban and suburban gardens in Western Australia. Three species have now darted off into the bush, where they sometimes form dense stands, crowding out the less ostentatiously colorful Australian natives that once occupied these sites.

Peter Day has learned how to fight the weed lily on behalf of native Australian plants. His recipe for *Watsonia* control includes application of Roundup to the densest stands. In areas with scattered intruders, he advocates digging out or cutting down the enemy one by one the old-fashioned way, with shovel or clippers. Then, just before the close of *Watsonia*'s reproductive season, he and any volunteers he can enlist make a sweep through infested areas to locate, remove, and destroy any and all seed-bearing flower stalks *before* the seed has been dropped to the ground. Because even one missed plant can deposit dozens of seeds, the plant police must be alert and keep at it year after year, a policy that Peter Day has followed. As a result, the area around Darlington is largely *Watsonia*-free, at least for the moment. The risk of reinvasion from areas outside the war zone means that others will eventually have to take on the work now done by Mr. Day, which is why it is a good thing that Australian organizations like the Environmental Weeds Action Network and Weed Busters have been formed to keep troops on the front line.

Although there have been some success stories when it comes to weedy exotics, the fungal root invader (*Phytophthora*) offers an even more substantial challenge. The darn thing is almost everywhere and yet cannot be seen hiding underground or inside the roots of its victims. The microscopic spores of the fungus can be picked up in mud attached to a vehicle, or even a hiking boot, and transported to new areas, there to multiply in the many vulnerable species that had hitherto been uninfected. Because of the ease with which the fungus can be spread and the devastating consequences of its introduction into what were once *Phytophthora*-free areas, the authorities have focused their efforts on trying to keep the fungus out of regions not yet infected. Among other things, CALM has installed covered trays with brushes at many trailheads, urging hikers to brush their boots into the tray before walking the trail lest they contribute to the spread of this insidious fungus. I have the feeling that this antidote to the spread of dieback is not likely to be highly effective, although it may raise general awareness of the extent of the problem. The outright closure of uncontaminated areas to walkers and vehicles makes more sense. This policy has been put into effect in remote parts of the Stirling Range and Fitzgerald River National Parks, two of the premier regions of botanical diversity in Western Australia.

If a shrub or tree does become infected, it can be saved, but to do so requires spraying with phosphoric acid, a chemical that increases the capacity of plants to resist fungal infection. Needless to say, this remedy is far too costly and labor-intensive to be applied to the millions upon millions of dieback-affected plants in southwestern Australia, although parts of the Stirling Range National Park are sprayed annually in recognition of the exceptional importance of this area for botanical diversity. This antidieback method has also been put to work in defense of an extremely rare and localized spider orchid, *Caladenia winfieldii*, that lives under tall eucalyptus trees. The orchid apparently resists *Phytophthora*, but the trees that provide critical shade and shelter for the orchid are susceptible to the fungus. By boosting the defenses of the protective eucalypts, which have also been declared off-limits to loggers, forest managers help keep the orchid alive. Incidentally, this orchid was so severely reduced in numbers (at one point, only seven flowering plants could be found in their little forest refugium) that they were failing to attract any wasp pollinators. CALM biologists had to step in to substitute for the missing wasps. By hand-pollinating the survivors over several years, Andrew Brown's crew increased the population of flowering plants to more than 170, not a huge figure on which to stake the survival of a species but better than seven.

So, yes, the dieback situation is depressing, but at least some partial solutions exist for this plant root killer. Likewise, the salinization problem is extremely serious, but here, too, a few small steps are being taken, which gives the optimists among us reason to remain hopeful. The trick in dealing with agriculture-induced increases in soil salts is to get the stuff away from plant roots, which one could in theory achieve by driving the salts far down in the soil beneath the plant root zone or by flushing the soil with water so that its salts were carried away in drains to be deposited elsewhere. Flushing salts out of soil is expensive because it costs so much to build the ditches and tiled drains needed to move salty water out of an affected region and into waterways where the salt will become someone else's problem. The ditch technique is favored by farmers in the western United States, who rely on Colorado River water in their irrigation schemes. The river is salty to begin with, which makes salt buildup in agricultural fields all the more inevitable. But by flushing agricultural fields in the Colorado River watershed, western agribusinesses send the salts downstream. By the time the unfortunate farmer in Mexico receives his share of the Colorado River, the fluid resembles sea water.

American salt flushing schemes are impractical for Australians, who lack a convenient Mexican dumping ground for their salt water. Instead, the Aussies have devised another possible remedy, which is to restore the tree and shrub cover to areas that have become salted up over the years.

The idea is that increased transpiration will lower the water table, driving the salts downward out of the root zone. Throughout the agricultural parts of southwestern Australia, one sees rows of salt-tolerant eucalypts and other hardy native trees along either side of degraded streams with their whitish soil encrustations and collection of long dead trees killed earlier by a rising water table. Although this remedy may slow the spread of salt across a field, the cure most certainly does not produce anything like a healthy-looking stream with native riparian vegetation. Among other things, the tree planters usually create completely unnatural-looking lines of eucalypts without any of the understory shrubs that abound in native Australian forests. The general rule in restoration ecology is that one can take something apart much more easily than one can put it back together again. Still, at least the authorities recognize the horrors of salinization and have enlisted farmers in an effort to slow the disassembly process.

Thus, were someone to develop a public television program on the wonderful plants and glorious natural ecosystems of Western Australia, the end of the hour might feature some words of muted optimism instead of unrelieved gloom about the prospects for these plants and places. Perhaps the most hopeful aspect is that many people really do care, with the result that a substantial park and reserve system has been established and well maintained in this part of the world. If I were employed by the Public Broadcasting System and in charge of the final scene for a nature film on orchids, I would take the viewer along to Cape Arid, one of the least visited of the Australian national parks, a huge block of land far out at the eastern edge of the wheat belt, where so little rain falls that few pioneer farmers set foot there, and those that did eventually cleared out before doing too much damage. After filming the vast Yokinup Bay with its fringe of pure white beach that runs for mile after mile in a great sweeping arc around to Mt. Arid on the far horizon, the camera team and I would head for the Len Otte trail, just a short walk up from the ocean's edge. The trail is a fine memorial for Mr. Otte, a deceased park ranger. We would wander along, first filming the luxuriant heath, which doesn't seem to be at all fazed by the shortage of precipitation, before moving into a woodland of small eucalypts with black and twisted trunks. I would have the cameraman sweep across the forest floor, littered with the pale brown leaves of the eucalypts, before slowly zooming in on a ruddy greenhood, yet another Western Australian species that has yet to be formally named. The orchid would be invisible initially, thanks

7.6
The ruddy greenhood, one of the many tiny terrestrial orchids that hide in the forests and heathland of Western Australia.

to its small size and muted coloration. But as the zoom lens brought the viewer closer, a three-inch-tall reddish brown plant would gradually materialize. As the flower took shape on the screen, viewers would see that the orchid appeared to have a cartoonish head with open mouth and a tongue-like labellum, spring-loaded and ready to leap up into the hood of the flower should the appropriate pollinating fly come darting in to land on the labellum.

The cameraman would then continue to pan from one ruddy greenhood to another, showing that an entire colony of these attractive plants was flourishing here, a colony almost certainly ignored by almost every person who comes to walk the Len Otte trail. When I stumbled across the first specimen I happened to see near this trail, I retraced my steps, finding a dozen more plants that I had walked right past even though I had been looking intently for orchids all the while. I find it reassuring to think that swarms of orchids must be hiding throughout Western Australia in the remaining protected natural areas, so easy to overlook, so inconspicuous, so strangely beautiful despite their small size. I take hope from the fact that of fifty-three species of Western Australian plants believed to be extinct in 1991, fully eighteen have been rediscovered in the field by amateurs and professional botanists who went looking for them in places that had not been surveyed adequately before. I want to imagine that other presumed extinct species and the many endangered plant species of Western Australia have refuges unknown to us currently, where they are hanging on in the face of habitat loss and competition from alien plants, each species a different story in the evolutionary book of life. It would be good to read all the chapters in this book at some point, but to do so will require that we do a better job at controlling our numbers and our appetites than we have done to date. This kind of message is implicit in a host of PBS nature programs but perhaps one more wouldn't hurt.

Let us imagine that although you are not necessarily an orchid obsessive, you would like to see some wild orchids, the better to appreciate the processes of evolution that have generated so many species, each with its own set of delightful adaptations. Perhaps you would enjoy a visit to Australia to search for flying duck orchids, as well as to admire the wombat, visit the Sydney Opera House, and climb Ayres Rock (or walk about its base, as suggested by the local aboriginal community, who consider walking up the rock a desecration of a sacred site). How should you go about the orchid hunt? There is no one right way to find Australian orchids. Much will depend on the time available to you and the cash at your disposal, but this is what I would do if I had two or three weeks for the trip and a functional credit card.

8

Happy

Hunting

After having flown into the country and had at least one night of rest as part of the painful process of biological clock resetting, I would pick up my rental car or campervan the next day, having made arrangements for the vehicle well beforehand. For many, the prospect of driving on the "wrong" side of the road is unnerving, but most North Americans have little trouble coming to grips with this aspect of Australian life. I, too, initially did not relish the challenge of driving Down Under, but by repeating "Think left" and "Keep left!" over and over, making these phrases a personal mantra, I managed to switch to the Australian mode of driving without too much trauma. But remember to think left whenever leaving a service station or parking area. If you forget, you may find yourself wondering why it is that an approaching truck or passenger car seems intent on crashing head-on into your vehicle, a thought that will be followed shortly by the realization that you, not the other driver, are in the wrong lane, a mistake that you had better correct in the next second or two.

Assuming that you do not have these unpleasant and disconcerting moments on the road, you will eventually arrive at your destination in good spirits, proud of your flexibility in vehicular management. To help you choose where to go, I recommend that you secure one of the free tourist booklets or papers that lists accommodations by town. These are widely available from car rental companies and tourist centers. If you are traveling in a rental car, make your way to a self-contained holiday unit (which will provide you with cooking facilities). If in a campervan, odds are you will head to a caravan park (which will provide you with a space to park your vehicle and an "ablution block" with toilets and showers). Almost every town, even the most insignificant ones, in Australia has its caravan park, which also provides caravans (trailers) for an evening rental. The price will be right, although in some places the conditions can be less than completely luxurious.

8.1

The cowslip orchid, *Caladenia flava*, one of the most abundant of orchids in Stirling Range National Park and elsewhere in southwestern Australia.

But which holiday unit destination or caravan park to choose? If your goal is to do some orchiding, pick a spot near an Australian national park or reserve. You will not find orchids of interest in a wheat field or pasture but must instead go somewhere with a decent patch of native bushland. The southwestern corner of Western Australia has a large number of parks and reserves, large and small, and at the right time of the year, these places will have at least some orchids on show. The Stirling Range National Park is the largest nature reserve in the southwest, with well over 250,000 acres and more than twelve hundred species of plants (more than 20 percent of the total count for the southwest). Among the botanical riches found in the park are more than 120 orchid species, greenhoods, flying ducks, hammer orchids, spider orchids, leek orchids, and more. Many of these species do especially well after an appropriately timed wildfire. Controlled burns are a regular feature of management policy in the Stirling Range (and elsewhere in southwestern Australia) as part of CALM's attempts to prevent the buildup of flammable materials that could sustain a massive, catastrophic, crown-burning, tree-destroying bushfire. Some persons feel that the controlled burns occur too often or at the wrong time of the year, with negative effects in terms of maintaining the full range of native flora and fauna in Australian woodlands. But

in some instances, CALM's burn-burn-burn schedule promotes flowering by certain orchids, because of the clearing of the undergrowth and release of nutrients from burnt plant material. The result is that orchid hunters in Australia often seek out woodlands that were scorched in the preceding year or two.

Because of CALM's fire promotion policy in the Stirlings, you may well come across large areas in the park that have been burnt in a preceding year and you will probably find these to be most rewarding. One can, in the right area in the right season, encounter astonishing numbers of certain species, dense pockets of twenty or thirty flying ducks instead the customary scattered singletons, long lines of purple enamel orchids glistening in the sunshine, carpets of yellow cowslips. Wear trousers and shirts that you are prepared to sacrifice, however, because pushing through the charcoaled stalks of small shrubs and brushing up against the blackened banksia trunks will leave your clothes much the worse for wear.

8.2

This handsome orchid found near Lake Poorarecup is a hybrid between *Caladenia polychroma* and *Caladenia caesarea*.

The park has a campground and is adjacent to several caravan parks, including one in the little town of Cranbrook to the west and the other near Chester Pass to the northeast. This latter caravan park, the Stirling Range Retreat, offers wildflower walks in season with an emphasis on the orchids for those who would like a little assistance in becoming acquainted with the local flora.

Just west of the Stirling Range, in the area between Cranbrook and Franklin, several small reserves have plenty of orchids in September and October. Let me recommend Lake Poorarecup Reserve with its semiprimitive camping area near a beach on the lake. Swimming and water skiing are things of the past now that this and other lakes in the area are known to harbor *Entamoeba histolytica*, an unpleasant protozoan that can infect

257

humans when water carries the bug up the nasal passages. Dysentery follows and, in rare cases, so does a cerebral abscess. Neither condition is desirable, so splashing in the lake is out. But so long as you stay clear of the water, you will find the woodlands bordering the lake richly rewarding, botanically speaking, with large and small spider orchids as well as leek, donkey, and dragon orchids.

I confess that I did have an unpleasant encounter at Lake Poorarecup with another group camping there. I met the men coming along a woodland track, followed by the women, who were carrying large bouquets of freshly picked spring wildflowers, including some handsome spider orchids. I expressed my dismay and the opinion that surely what they had done was illegal (which it was). The matriarch of the group told me that she had been coming to the area to pick wildflowers for forty years and was not about to stop. I did not reply that forty years of wildflower destruction was nothing to brag about, but I permitted myself the thought. No doubt the orchid pickers went to work with a vengeance after we left, so that by playing the native plants policeman I probably did more harm than good. But with luck you will not encounter plant pickers at the lake and will instead delight in the opportunities to look at and photograph (but not destroy) the native orchids of the area.

Lake Poorarecup and the Stirling Range lie to the southeast of Perth; they can be most profitably visited between September and November. Other orchid-rich areas occur due north of the big city. On your way to these northern sites, let me recommend a stop at Western Flora Caravan Park, near Eneabba and just off the Brand Highway, the main coastal route to the north. Western Flora is run by Allan and Lorraine Tinker,

8.3

An unusual white-flowered donkey orchid discovered only recently by Allan Tinker, owner of Western Flora Caravan Park. This species, like many other Western Australian orchids, has not yet received a formal scientific name.

two remarkably friendly and welcoming people. Allan is also a self-made botanist and conservationist who has had the immense pleasure of finding undescribed orchids and other plants during those rare times when he can slip away from caravan park management and upkeep. Visitors to the park will find his affection and enthusiasm for botany contagious, especially during his evening walking tours of the native bush on his property. Allan also sometimes leads driving tours to botanically rich locations nearby, and he generously directs the independent searcher to other attractive places.

Western Flora lies in the northern sandplains district, with its extraordinary shrub diversity. The northernmost extension of this botanically

unusual habitat is Kalbarri National Park, which, like the Stirling Range, offers especially exciting orchid hunting opportunities in season. Because of the geographically isolated nature of the Kalbarri sandplains, any number of orchid species are found here and nowhere else, among them the smooth-lipped flying duck orchid mentioned earlier in the book as well as the kneeling hammer orchid. These two species occur in several scattered populations along the Murchison River that passes through the park to the ocean. You will be fortunate to see either the local flying duck or hammer orchid, but they do grow near one of the main park destinations, the Z-bend lookout at the end of a sometimes heavily corrugated dirt road. Car camping is not permitted within the park, but the town of Kalbarri is well appointed with motels, cabins, and caravan parks, so that lodging should not be a problem (except perhaps during school holidays). The best time for orchids in this area is between August and mid-September.

Although a few orchid species occur north of Kalbarri, the park is to all intents and purposes the northern border of Western Australia's terrestrial orchid range; Cape Arid National Park, another wonderful destination (see Chapter 7), anchors the southeastern end of orchid country. And I have not yet mentioned the Margaret River-Augusta area on the far southwestern coast, with its vineyards, upscale eateries, and excellent orchid habitats, or Walpole-Nornalup National Park, with its giant eucalypts, extensive network of hiking trails, and full spectrum of orchids. Even Kings Park right in Perth is well worth visiting, especially if you are there during the peak orchid period between August and early October, when cowslips, purple enamels, and sandplain spider orchids share the forest floor with a rich variety of flowering shrubs.

As noted, the season of your visit will have much to say about the success you have in tracking down orchids. Nor surprisingly, different species have different flowering periods, and if you arrive too early or too

8.4
The kneeling hammer orchid, *Drakaea concolor*, a species considered endangered because it occurs in just a handful of sites, mostly in Kalbarri National Park.

late, you will be as disappointed as Darwin was during his visit to Western Australia. Although the Antipodean spring (September and October) is the best overall for southwestern Australia, August and early November offer good orchid hunting as well, and some species bloom in every month of the year (but February and March offer very little).

Given the seasonality and diversity of the orchids, a good field guide is obviously essential. As you know by now, the essential field guide for southwestern Australia's orchids is Noel Hoffman and Andrew Brown's *The Orchids of South-West Australia*. This book provides the vital data on the time of flowering and the appearance of the different species and is, in 2007, to be replaced with a new guide illustrating all four hundred or so Western Australia orchids, including those found in the far northwestern part of the state. The new book, which is being written by Andrew Brown, Kingsley Dixon, and Steve Hopper, will contain whole plant paintings of each species, with multiple illustrations of those that are variable in color, size, or height. As I have mentioned, some species are so similar in appearance that discriminating among them is a challenge, but perhaps not an insuperable one once the new field guide is available. Fortunately, a large proportion of Western Australian orchids are sufficiently distinctive that the currently available orchid book will enable you to put a name on most of the species you encounter.

8.5

The orchids of Tasmania features an assortment of greenhoods, one of which is shown on the left, and donkey orchids, one of which is shown on the right.

In addition, Western Australia has two organizations devoted to their magnificent wildflowers. Persons interested in the orchids and other plants of this state may wish to contact the Wildflower Society of Western Australia (Inc.), P.O. Box 64, Nedlands WA 6009 and/or Western Australia Native Orchid Study & Conservation Group (Inc.) at P.O. Box 323, Victoria Park WA 6100. The WANOSCG Web site is at http://communities.ninmsn.com.au/wanoscg. The Australian Native Orchid Society also has a Web site, http://www.anos.org.au/, that may be of interest should you wish to learn more about orchids throughout Australia.

In this regard, note that small terrestrial orchids are not limited to Western Australia and neither are field guides to orchid identification. *The Orchids of South Australia* by R. J. Bates and J. Z. Weber, *The Orchids of Victoria* by Gary Backhouse and Jeffrey Jeanes, *The Orchids of New South Wales and Victoria* by Tony Bishop, and *The Orchids of Tasmania*

by David Jones and coauthors are all good to excellent field guides for orchids found in the southern tier of Australian states. Having spent some time in Tasmania, I can report that the orchid hunter's strategy of studying the field guide and visiting national parks and reserves works just as well there as in Western Australia. Tasmania is sufficiently removed from Western Australia so that the floras of the two states are almost entirely different, which means I was able to secure a rich harvest of novel species as a first-time visitor to the area. I am sure that the same rule applies to the mainland states, including Queensland, for which, unfortunately, no field guide seems to be available.

Should you wish to combine a trip to New Zealand with your Australian adventure, you will be able to observe and admire a variety of small terrestrial orchids, many of which belong to genera that also occur in Australia. And you can make use of *The Nature Guide to New Zealand Native Orchids* by Ian St. George and others to put a name on the species you find (there are more than a hundred to look for in New Zealand).

Now it is true that travel to Australia (or New Zealand) requires a considerable investment of time and energy, which may mean that you cannot mount a foreign expedition in search of orchids. Nevertheless, you may be willing to look for a terrestrial orchid or two in a natural setting near your home as opposed to gawking at exotic greenhouse orchids or the potted specimens at Trader Joe's or Home Depot, where handsome mass-produced orchids are often sold. The native orchid hunter who lives in the United States will have around two hundred species to go after. If you are a Canadian, rejoice in the seventy-five orchids that grow in Canada. Most individual states and provinces have a couple of dozen species, some more, some less. My home state of Arizona is about aver-

8.6

One of the many native orchids of New Zealand, *Caladenia lyallii.*

age, with twenty-six native species, and Florida is tops, with about 110. Hawaii, however, comes in with only three species.

Some persons have crisscrossed our continent hunting for orchids to admire and photograph. One such dedicated individual is Philip Keenan, whose *Wild Orchids across North America* chronicles his mission to track down as many species as possible. He enjoyed much success, in large part because of his many contacts across the United States and Canada who were willing to lead him to rare and local species that he would have been unlikely to find quickly on his own.

Reading Keenan's book may stimulate you to get out and get going from Alaska to Arizona, California to Ontario, with a camera in hand.

However, because you and I are unlikely to have cultivated cross-continental acquaintances familiar with the unusual orchids in their neighborhood, we may have to settle for more limited trips in which any native orchids encountered provide an incidental bonus for the hiker and photographer. During my travels in Arizona I have not been overwhelmed with orchid sightings. But fortunately, my parents live in northern Virginia close to forested hills and mountains where the yellow lady's slipper (see Figure 1.7) blooms in the late spring. Although yellow lady's slippers are relatively common and extremely widespread, the orchid is nonetheless beautiful, a show stopper. In the woods where my father and I found the plant, many others had preceded us, as revealed by the trampled trackways encircling the flowering specimens. The many observers had obviously fluttered in like moths attracted to a light bulb. My dad and I, too, marched around the plants, examining them closely, our pleasure magnified by knowing (thanks to Darwin) how the slipper works to trap pollinators and then guide them past the column so that the bees can do their job.

8.7
North America is home to many fewer orchids than Australia, but nonetheless, the continent is blessed with some lovely species, such as the showy orchis and the calypso orchid, two widespread North American orchids.

Although the lady's slippers have orchid written all over them, thanks to their stunning and familiar flowers, other North American members of the family are far less well known and generally more modest in appearance. When my father and I were coming back from our combination orchid and bird walk though the Dick Thompson Conservation Area in the foothills of the Blue Ridge Mountains of Virginia, I happened to look down at the right moment and catch a glimpse of white, purple, and green that made me freeze in midstride. The gestalt of what proved to be a showy orchis was so similar to that provided by some of the Australian species that I knew at once I was onto an orchid, even though I had to get out *A Field Guide to the Flowers: Northeastern and North-Central North America* by Roger Tory Peterson and Margaret McKenny in order to put a name on my Virginian find.

Giving an orchid one has discovered its proper name adds greatly to one's enjoyment of the moment, as I have mentioned before. Fortunately, in North America, as well as in Australia, the would-be amateur orchid hunter has any number of guides to identification, ranging from books that cover many flowering plants (there are several in the Peterson field

guide series that cover different regions) to those that specifically focus on orchids. Philip Keenan recommends Carl Luer's *The Native Orchids of the United States and Canada*, which deals with everything except the Floridian species, which Luer covered in another volume (both of which are out of print now). The North American orchids are also covered by Donovan Correll in his *Orchids of North America North of Mexico*, and by a much newer book by Paul Martin Brown (*Wild Orchids of North America, North of Mexico*), published in 2003. As an Arizonan, I am entirely content with Ronald A. Coleman's *The Wild Orchids of Arizona and New Mexico*. If I lived in California, I would be sure to own Coleman's *The Wild Orchids of California*. If I lived in the southeast, Stanley Bentley's *Native Orchids of the Southern Appalachian Mountains* would be in my bookcase.

An owner of any of these books would not only be able to identify the species encountered, he or she would also be able to examine the catalogue of orchid illustrations in the books to form the appropriate search images for later field work. It helps immensely to have an idea of what might be around the corner in order to detect the plant when it actually appears in one's visual field. I have no doubt that because I had previously seen photographs of the calypso orchid, I was better prepared to detect the plant when I was hiking through a forest near Mount Rainier. There, poking up through the leaf litter, was first one, then another, and then still more specimens of an orchid whose appearance was familiar to me, even though I had not seen one in the wild before. The orchid would be entirely at home in Western Australia, where so many species are only a few inches tall. The calypsos I found were small and inconspicuous except for the orchid's colorful flower, whose great complexity required close-range inspection. Because I am nearsighted, I stretched out right on the ground next to the plant and took off my glasses, the better to absorb the elegant design of the calypso flower.

The field guides also help by providing information on the flowering seasons of the orchids that one would like to find. Although the leaves of some species are attractive, it is of course the blossoms that generate the real excitement. Because most species can be found in flower for only a month or two, it's good to know when those months are if you are to have a real chance at the adrenalin rush that comes from finding a flowering orchid that is new for your list.

Of course, the ultimate in an adrenalin high would be the discovery of an entirely new species of orchid. Just forget about it, however, if you are going to do your hunting only in the United States or Canada. Not only are there relatively few orchids in these countries, but with very few exceptions, those present were found and named long ago. No, if you want a new species of your own, the United States is out, and to all intents and purposes, so is Australia, although, as I mentioned earlier, professional botanists are still naming new species to this country's already extensive list. But even though there could be some more species still waiting for a sufficiently dedicated searcher in Oz, the odds of finding them are not good for a pure amateur given the intensity with which that country has been examined by trained botanists. Nevertheless, as you may recall, my friends, my wife, and I did encounter a flying duck orchid at Cape Arid in November 2001 that had been found only a few years previously, so recently that it had not yet made its way into Hoffman and Brown's orchid field guide.

And in October 2003 I had another experience with flying ducks that led me to think that I might just possibly have come up with something altogether new. In this instance, my wife and I were some 500 kilometers to the north of Perth in Kalbarri National Park, following our own advice to take advantage of the botanical riches of this park. We were after *Paracaleana lyonsii*, the smallest of all the flying duck orchids and the only one whose plants can have more than two flowers. The species was discovered a mere decade ago, in part because of its fairly restricted geographic range but even more so because of its small size and habit of hiding in the shade of dense patches of dark green sedges. I had hunted for the midget flying duck on several other trips to the area without success and had decided that the only way I was going to see this remarkable species was to enlist the aid of someone who knew exactly when and where to find the thing. To this end, I called Margaret Quicke at Eurardy Station just north of the park to make arrangements to go on a wildflower tour; Margaret and her husband, Bruce, have a sheep and wheat farm but do a little ecotourism on the side, leading visitors over the tracks on their property that pass through wonderful stands of flowering plants in season. During my phone call, my hopes of seeing *P. lyonsii* received a serious setback when Margaret told me that she had seen this flying duck just once in seventeen years of searching. Nevertheless I forged ahead and booked a tour of Eurardy.

On the day we drove to the station, I decided to stop along the main highway just south of Eurardy in an area that met the orchid book's description of the habitat preferred by the midget flying duck. The yellow sandy soil supported many patches of sedges as well as a host of grevilleas, acacias, banksias, little eucalyptus trees, and other elegant plants. I wandered among this attractive community for some time before seeing a little stick of a plant, two inches high, with a couple of buds poking up in a small bare patch surrounded by a ring of sedges. Even though the plant was not yet in flower there was no doubt that I was looking at the midget flying duck. I dashed off to locate my wife and bring her back for a serious search in the vicinity of my discovery, a good decision because Sue found a flowering specimen in about ten minutes. I saluted my wife and enjoyed again the wonderful sensations associated with orchid discovery. Then I set about securing a photograph of the plant. After about five minutes of struggling with my digital camera, whose operating complexities I still have not completely mastered, I realized that there were two other midget flying ducks less than a foot away from the specimen that my wife had found and that I had been fixated on. The small size and pale sandy color of the plants helped them blend into the background so well that I could easily see why Hoffman and Brown tell their readers that this species is a challenge to find. Which made it all the more exciting that we had done so.

It was then that I saw with considerable astonishment that the flock of three midget flying ducks were accompanied by a fourth individual, a flying duck that was most definitely not *P. lyonsii*. Instead, this species was a much larger, much darker *Paracaleana*. I knew that Kalbarri was home to one species of this sort, *P. terminalis*, but I was quite sure that what I was looking at had to be different. The black calli of "my" species covered the outer tip of the labellum, not just the very outermost point. Furthermore, the flowering season of *P. terminalis* had been over for almost a month, according to Hoffman and Brown's field guide. So here was another species altogether. I permitted myself to imagine what it would be like to present photographs of this orchid to an amazed and admiring convocation of orchid botanists.

When Sue and I found more specimens of "my" species, I photographed a whole series of individuals. Indeed, this flying duck was so common in the area that I felt a little deflated. Surely other persons had found the thing before. And in fact, when I reached Eurardy Station and had a chat with Margaret Quicke, she told me that on that very day a team from the Western Australian Herbarium had found the species in question. She told me that the orchid was *P. terminalis*, the very species that I had ruled out on the basis of its too ample calli.

I stored my battery of photographs on my computer and the happy memories of discovery in my head. A considerable time later I contacted Andrew Brown, thinking that he would know of the flying duck and would be able to tell me its name. (I had independently decided that it appeared to be either *Paracaleana triens* or a very close relative of this species, whose nearest neighboring population is, however, hundreds of kilometers to the south.) Andrew graciously agreed to take a look at my photographs. When I took them to him, I was hoping against hope that he might declare "my" flying duck a previously unknown species similar to but distinct from *P. triens*, a declaration that would enable me to claim codiscovery of my very own orchid species. However, when Andrew examined the first of my photos thoughtfully, he named the orchid *P. terminalis*, which I then learned exhibits a fair amount of variation in the extent to which the duck's beak is covered in calli. Moreover, some populations flower into October, contrary to the information contained in the field guide. Thus, I appeared to have been far too eager to locate a new species than was justified by the facts, a conclusion that I found less than exhilarating.

8.8
A new species of flying duck orchid.

But as Andrew went through the rest of the photographs I had presented him with, he pointed out that some of the individuals possessed a strangely humped labellum. The hump was unlike that of *P. nigrita*, the common flying duck, which in any event does not occur anywhere near Kalbarri. At the time when my wife and I had been searching for flying ducks at the midget flying duck site, I had in fact noticed that some specimens possessed a conspicuous hump two-thirds of the way toward the tip of the labellum. But I had assumed that these plants and the ones with smoother, flatter labella were all members of the same species, albeit a variable one. Andrew, however, felt that the humped specimens possibly represented a third species in addition to *P. terminalis* and *P. lyonsii*. He said he would have to visit the location in 2004 to check things in person but, as a result of our chat, my hope that I had actually found a novel species after all remained alive. (Note added later: Andrew has confirmed that "my" flying duck orchid is indeed a new species, a most exciting event for me.)

If you wish to come up with something totally new in the way of an orchid, you will probably have to be more adventurous than I have been. Before I realized that I had actually found a new species of orchid, I wrote to my Ph.D. advisor, the ornithologist Ernst Mayr, who went off to New Guinea in 1928 in search of birds of paradise in five major coastal

mountain ranges and brought back nearly forty species of orchids new to science. At the time Mayr was twenty-three and had never traveled outside of Europe. To enter the wilderness in New Guinea in 1928 was to run a real risk of being speared or clubbed by the often less than hospitable Stone Age tribesmen who inhabited these mountains. Alternatively, Mayr could have died from any of the many tropical diseases endemic to New Guinea, of which cerebral malaria is only one unpleasant example. However, Mayr not only survived his one-man expedition, despite many opportunities to do otherwise, he also went on to become one of the most influential evolutionary biologists of all time. His field work in New Guinea helped shape his ideas on evolution, and the immediate products of the expedition were the large collections of birds he brought back. The orchids were a minor sideline. Nevertheless, when I learned of Mayr's incidental botanical triumphs, I wrote him, expressing a certain amount of envy at his orchid successes. Mayr was well into his nineties at the time, but he wrote to offer me the following advice, which can apply to all who suffer from a desire to find an unnamed species of orchid:

Dear John,

It is quite clear to me that you will die of your urge to discover a new species of orchid unless something is done about it. So, here is my suggestion on how to reach your goal!

First of all, forget Australia. Too many people have combed all the suitable areas. Then what? Of course you must go to New Guinea.

Second, get a species list and mark on a New Guinea map all the type localities of the named orchid species. You will discover that there are no type localities on quite a few major and minor mountain ranges. And every range has endemic [unique] species.

Third, select the most easily accessible range.

Fourth, most New Guinea orchids are epiphytes that grow way up in the canopy. You cannot possibly get up there. But the canopy is very obliging, it will come to you. Everywhere on the forest floor are broken, fallen branches or whole trees that have succumbed to old age. They are always covered with epiphytes. Of course you do not know which are new but you collect them all. The more insignificant they look, the more likely they are to be a new species.

Fifth, your only problem will be to dry the plant specimens in their herbarium sheets. I did it over the fire in smoky native huts. Lots of fun!

Happy hunting,

Ernst

Happy hunting indeed. May my readers regularly experience the enjoyment that comes from finding orchids, whether on your home ground or in a foreign land. When a find is made, I hope that you will also derive pleasure from knowing something about the evolutionary biology of this group of plants. You can consider yourselves, if you wish, part of a small club whose founding member was Charles Darwin, who provided naturalists with the ability to look at things through an evolutionary lens. The view through this lens adds immensely to an appreciation of all living things, orchids included. This extraordinary group offers so much more than the raw material for a fast-fading corsage or a status-enhancing greenhouse display or a medieval source of Viagra. A flying duck orchid may be neither edible nor good-looking, but it has evolved a nifty way of getting pollinated, which it does in a world hidden to most people. This is a plant worth celebrating, worth getting excited about, a plant that goes beyond the merely aesthetic or the crassly commercial to teach us something that cannot be learned from the vanilla orchid, the rose, the potato, or any other economically valuable plant. And remember that the ten or so flying ducks are part of a larger assemblage of tens of thousands of species, each orchid seemingly trying to outdo the others in terms of strangeness of form and intricacy of ecological interactions. With flowers like that, who cares if we have yet to find out how to earn a dollar or two from these plants, which are totally useless and valuable beyond measure, a conclusion that I am sure Darwin would second wholeheartedly.

References

For some basic books on various aspects of orchid and plant biology, I can recommend the following, two of which provide historical context (Allan and Darwin), while one offers a modern academic review of orchidology (Arditti). The book by Hoffman and Brown focuses on the orchids of southwestern Australia, and the remaining two on the list are good reading for a general audience, whether interested specifically in orchids (van der Pijl & Dodson) or the plant world as a whole (Bernhardt).

Allan, M. 1977. *Darwin and His Flowers*. New York: Taplinger.

Arditti, J. 1992. *Fundamentals of Orchid Biology*. New York: Wiley.

Bernhardt, P. 1999. *The Rose's Kiss: A Natural History of Flowers*. Chicago: University of Chicago Press.

Darwin, C. 1892. *The Various Contrivances by which Orchids are Fertilised by Insects*. New York: D. Appleton.

Hoffman, N. & Brown, A. 1998. *The Orchids of South-West Australia*, 2nd ed. Nedlands: University of Western Australia Press.

van der Pijl, L. & Dodson, C. H. 1966. *Orchid Flowers, Their Pollination and Evolution*. Coral Gables, FL: University of Miami Press.

The following chapter-by-chapter lists contain many papers in moderately obscure biological journals that I referred to when writing the chapter in question. Let me acknowledge my debt to Aaron Hicks of the School of Life Sciences at Arizona State University for his help in tracking down some of the relevant literature listed below. If you wish to find these papers, you will probably need access to a good university library, although some of the books and journals mentioned here will be available at your local town or city library.

Chapter 1: Warty Hammer Orchids, Adaptations, and Darwin

Bierce, A. *The Devil's Dictionary*. New York: Dover Publications.

Coen, E. 2002. The making of a flower. *Natural History*, 111, 48–55.

Dennett, D. C. 1995. *Darwin's Dangerous Idea*. New York: Simon and Schuster.

Gould, S. J. 1986. Cardboard Darwinism. *New York Review of Books*, 33, 47–50.

Gould, S. J. 1997. The pleasures of pluralism. *New York Review of Books*, 44, 47–52.

Gould, S. J. & Lewontin, R. C. 1979. The spandrels of San Marco and the Panglossian paradigm: A critique of the adaptationist programme. *Proceedings of the Royal Society of London B*, 205, 581–598.

Johnson, S. D. & Edwards, T. J. 2000. The structure and function of orchid pollinaria. *Plant Systematics and Evolution*, 222, 243–269.

Lawler, L. J. 1984. Ethnobotany of the Orchidaceae. In *Orchid Biology, Reviews and Perspectives, III* (Ed. by Arditti, J.). Ithaca, NY: Cornell University Press.

Peakall, R. 1990. Responses of male *Zaspilothynnus trilobatus* Turner wasps to females and the sexually deceptive orchid it pollinates. *Functional Ecology*, 4, 159–167.

Sargent, R. D. 2004. Floral symmetry affects speciation rates in angiosperms. *Proceedings of the Royal Society of London Series B*, 271, 603–608.

Stoutamire, W. P. 1974. Australian terrestrial orchids, thynnid wasps and pseudocopulation. *American Orchid Society Bulletin*, 43, 13–18.

Stoutamire, W. P. 1983. Wasp-pollinated species of *Caladenia* (Orchidaceae) in southwestern Australia. *Australian Journal of Botany*, 31, 383–394.

Wallace, L. E. 2003. The cost of inbreeding in *Platanthera leucophaea* (Orchidaceae). *American Journal of Botany*, 90, 235-242.

Chapter 2: The Adaptations of Behaving Plants

Becerra, J. X. 1994. Squirt-gun defense in *Bursera* and the chrysomelid counterploy. *Ecology*, 75, 1991–1996.

Bernhardt, P. 1995. Observations on the floral biology of *Pterostylis curta* (Orchidaceae). *Cunninghamia*, 5, 1–8.

Darwin, C. 1893. *The Movements and Habits of Climbing Plants.* New York: D. Appleton.

Darwin, C. 1896. *Insectivorous Plants.* New York: D. Appleton.

Darwin, C. 1896. *The Power of Movement in Plants.* New York: D. Appleton.

Ebel, F. 1974. Beobachtungen über das Bewegungsverhalten des Pollinariums von *Catasetum fimbriatum* Lindl. Während Abschuß, Flug und Landung. *Flora*, 163, 342-356.

Evans, J. P. & Cain, M. L. 1995. A spatially explicit test of foraging behavior in a clonal plant. *Ecology*, 76, 1147–1155.

Ganeshaiah, K. N. & Shaanker, R. U. 1993. Foraging decisions by plants: Making a case for plant ethology. *Current Science*, 65, 371–373.

Hill, B. S. & Findlay, G. P. 1981. The power of movement in plants: The role of osmotic machines. *Quarterly Reviews of Biophysics*, 14, 173–222.

Juniper, B. E., Robins, R. J. & Joel, D. M. 1989. *The Carnivorous Plants*. London: Academic Press.

Kelly, C. K. 1990. Plant foraging: A marginal value model and coiling response in *Cuscuta subinclusa*. *Ecology*, 71, 1916–1925.

Kelly, C. K. 1992. Resource choice in *Cuscuta europaea*. *Proceedings of the National Academy of Sciences*, 89, 12914–12917.

Koller, D. 2000. Plants in search of sunlight. *Advances in Botanical Research*, 33, 35–131.

Legendre, L. 2000. The genus *Pinguicula* L. (Lentibulariaceae): An overview. *Acta Botanica Gallica*, 147, 77–95.

Leopold, A. C. & Jaffe, M. J. 2000. Many modes of movement. *Science*, 288, 2131–2132.

Malone, M. 1996. Rapid, long-distance signal transmission in higher plants. *Advances in Botanical Research*, 22, 163–228.

Muraoka, H., Takenaka, A., Tang, R., Koizumi, H. & Washitani, I. 1998. Flexible leaf orientations of *Arisaema heterophyllum* maximize light capture in a forest understory and avoid excess irradiance at a deforested site. *Annals of Botany*, 82, 297–307.

Nilsen, E. T. 1992. Thermonastic leaf movements: a synthesis of research with *Rhododendron*. *Botanical Journal of the Linnaean Society*, 110, 205–233.

Romero, G. A. & Nelson, C. E. 1986. Sexual dimorphism in *Catasetum* orchids: Forcible pollen emplacement and male flower competition. *Science*, 232, 1538–1540.

Silvertown, J. & Gordon, D. M. 1989. A framework for plant behavior. *Annual Review of Ecology and Systematics*, 20, 349–366.

Thompson, J. N. 1981. Reversed animal-plant interactions: The evolution of insectivorous and ant-fed plants. *Biological Journal of the Linnean Society*, 16, 147–155.

Williams, S. E. 1976. Comparative sensory physiology of Droseraceae: Evolution of a plant sensory system. *Proceedings of the American Philosophical Society*, 120, 187–204.

Williams, S. E. & Bennett, A. B. 1982. Leaf closure in the Venus flytrap: An acid growth-response. *Science*, 218, 1120–1122.

Zamora, R. 1990. Observational and experimental study of a carnivorous plant-ant kleptobiotic interaction. *Oikos*, 59, 368–372.

Chapter 3: Adaptations and Maladaptations

Alcock, J. & D. T. Gwynne. 1987. Courtship feeding and mate choice in thynnine wasps (Hymenoptera: Tiphiidae). *Australian Journal of Zoology*, 35, 451–459.

Ayasse, M., Paxton, R. J. & Tengo, J. 2001. Mating behavior and chemical communication in the order Hymenoptera. *Annual Review of Entomology*, 46, 31–78.

Berliocchi, L. 2000. *The Orchid in Lore and Legend*. Portland, OR: Timber Press.

Borg-Karlson, A.-K. 1990. Chemical and ethological studies of pollination in the genus *Ophrys* (Orchidaceae). *Phytochemistry*, 29, 1359–1387.

Bower, C. C. 1996. Demonstration of pollinator-mediated reproductive isolation in sexually deceptive species of *Chiloglottis* (Orchidaceae: Caladeniinae). *Australian Journal of Botany*, 44, 15–33.

Cain, A. J. 1964. The perfection of animals. In *Viewpoints in Biology*, vol. 3 (Ed. by Carthy, J. D. & Duddington, C. L.), pp. 36–63. London: Butterworths.

Dominy, N. J. & Lucas, P. W. 2001. Ecological importance of trichromatic vision to primates. *Nature*, 410, 363–366.

Duchaine, B., Cosmides, L. & Tooby, J. 2001. Evolutionary psychology and the brain. *Current Opinion in Neurobiology*, 11, 225–230.

Gaulin, S. J. C. & McBurney, D. H. 2001. *Psychology, An Evolutionary Approach*. Upper Saddle River, NJ: Prentice-Hall.

Gauthier, I., Skudlarski, P., Gore, J. C., & Anderson, A. W. 2000. Expertise for cars and birds recruits brain areas involved in face recognition. *Nature Neuroscience*, 3, 191–197.

Geary, D. C. 1998. *Male, Female: The Evolution of Human Sex Differences*. Washington, DC: American Psychological Association.

Hansen, E. 2000. *Orchid Fever: A Horticultural Tale of Love, Lust and Lunacy*. New York: Pantheon Books.

Jones, B. C., Little, A. C., Penton-Voak, I. S., Tiddeman, B. P., Burt, D. M. & Perrett, D. I. 2001. Facial symmetry and judgements of apparent health: Support for a "good genes" explanation of the attractiveness-symmetry relationship. *Evolution and Human Behavior*, 22, 417–429.

Kaiser, R. 1993. *The Scent of Orchids: Olfactory and Chemical Investigations*. Amsterdam: Elsevier Scientific.

Mant, J., Schiestl, F. P., Peakall, R., & Weston, P. H. 2002. A phylogenetic study of pollinator conservatism among sexually deceptive orchids. *Evolution*, 56, 888–898.

Rikowski, A. & Grammar, K. 1999. Human body odour, symmetry and attractiveness. *Proceedings of the Royal Society of London B*, 266, 869–874.

Schieb, J. E., Gangestad, S. W. & Thornhill, R. 1999. Facial attractiveness, symmetry and cues of good genes. *Proceedings of the Royal Society of London B*, 266, 1913–1917.

Schiestl, F. P. & Ayasse, M. 2001. Post-pollination emission of a repellent compound in a sexually deceptive orchid: A new mechanism for maximizing reproductive success? *Oecologia*, 126, 531–534.

Schiestl, F. P., Peakall, R., Mant, J. G., Ibarra, F., Schultz, C., Franke, S., & Wittko, F. 2003. The chemistry of sexual deception in an orchid-wasp pollination system. *Science*, 302, 437–438.

Singer, R. B. 2002. The pollination mechanism in *Trigonidium obtusum* Lindl (Orchidaceae: Maxillariinae): Sexual mimicry and trap flowers. *Annals of Botany*, 89, 157–163.

Wong, B. B. M. & Schiestl, F. P. 2002. How an orchid harms its pollinator. *Proceedings of the Royal Society of London B*, 269, 1529–1532.

Chapter 4: The History in Evolution

Armstrong, P. 1985. *Charles Darwin in Western Australia: A Young Scientist's Perception of an Environment.* Nedlands: University of Western Australia Press.

Ayasse, M., Schiestl, F. P., Paulus, H. F., Löfstedt, C., Hansson, B., Ibarra, F. & Francke, W. 2000. Evolution of reproductive strategies in the sexually deceptive orchid *Ophrys sphegodes*: How does flower-specific variation of odor signals influence reproductive success? *Evolution*, 54, 1995–2006.

Behe, M. 1996. *Darwin's Black Box: The Biochemical Challenge to Evolution.* New York: Free Press.

Cameron, K. M., Chase, M. W., Whitten, W. M., Kores, P. J., Jarrell, D. C., Albert, V., Tukawa, T., Hills, H. G. & Goldman, D. H. 1999. A phylogenetic analysis of the Orchidaceae: Evidence from *rbcL* nucleotide sequences. *American Journal of Botany,* 86, 208–224.

Cameron, K. M., Wurdack, K. J. & Jobson, R. W. 2002. Molecular evidence for the common origin of snap-traps among carnivorous plants. *American Journal of Botany,* 89, 1503–1509.

Chapman, G. 1997. Orchids: a witness to the Creator. (answersingenesis. org)

Darwin, C. 1871. *The Descent of Man, and Selection in Relation to Sex.* London: John Murray.

Dawkins, R. 1996. *The Blind Watchmaker.* New York: Norton.

Dembski, W. A. 1998. *The Design Inference: Eliminating Chance through Small Probabilities.* Cambridge, UK: Cambridge University Press.

Desmond, A. & J. Moore. 1991. *Darwin: The Life of a Tormented Evolutionist.* New York: Norton.

Futuyma, D. J. 1997. Miracles and molecules. *Boston Review,* 22 (Feb./ Mar.), 29–30.

Gellon, G. & McGinnis, W. 1998. Shaping animal body plans in development and evolution by modulation of *Hox* expression patterns. *Bioessays,* 20, 116–125.

Ghiselin, M. T. 1984. Foreword. In *The Various Contrivances by which Orchids are Fertilised by Insects.* Chicago: University of Chicago Press.

Gigord, L. B. D., Macnair, M. R., Stritesky, M. & Smithson, A. 2002. The potential for floral mimicry in rewardless orchids: An experimental study. *Proceedings of the Royal Society of London B,* 269, 1389–1395.

Gould, S. J. & Vrba, E. S. 1982. Exaptation: A missing term in the science of form. *Paleobiology*, 8, 4–15.

Johnson, P. E. 1997. *Defeating Darwinism by Opening Minds*. Downers Grove, IL: InterVarsity Press.

Johnson, S. D. 2000. Batesian mimicry in the non-rewarding orchid *Disa pulchra*, and its consequences for pollinator behavior. *Biological Journal of the Linnean Society*, 71, 119–123.

Johnson, S. D. & Nilsson, L. A. 1999. Pollen carryover, geitonogamy, and the evolution of deceptive pollination systems in orchids. *Ecology*, 80, 2607–2619.

Johnson, S. D., Peter, C. I. & Ågren, J. 2004. The effects of nectar addition on pollen removal and geitonogamy in the non-rewarding orchid *Anacamptis morio*. *Proceedings of the Royal Society of London B*, 272, 803–809.

Johnson, S. D. & Steiner, K. E. 1997. Long-tongued fly pollination and evolution of floral spur length in the *Disa draconis* complex. *Evolution*, 51, 45–53.

Johnson, S. D., Steiner, K. E., Whitehead, V. B. & Vogelpoel, L. 1998. Pollination ecology and maintenance of species integrity in co-occurring *Disa racemosa* L.f. and *Disa venosa* Sw. (Orchidaceae) in South Africa. *Annals of the Missouri Botanical Garden*, 85, 231–241.

Keynes, R. 2001. *Annie's Box: Charles Darwin, His Daughter and Human Evolution*. London: Fourth Estate.

Nilsson, L. A. 1992. Orchid pollination biology. *Trends in Ecology and Evolution*, 7, 255–259.

Norell, M., Ji, Q., Gao, K. Q., Yuan, C. X., Zhao, Y. B. & Wang, L. X. 2002. "Modern" feathers on a non-avian dinosaur. *Nature*, 416, 36–37.

Paxton, R. J. & Tengö, J. 2001. Doubly duped males: the sweet and sour of the orchid's bouquet. *Trends in Ecology and Evolution*, 16, 167–169.

Peakall, R. & Beattie, A. J. 1996. Ecological and genetic consequences of pollination by sexual deception in the orchid *Caladenia tentactulata*. *Evolution*, 50, 2207–2220.

Pichersky, E. & Gershenzon, J. 2002. The formation and function of plant volatiles: Perfumes for pollinator attraction and defense. *Current Opinion in Plant Biology*, 5, 237–243.

Reeve, H. K. & Sherman, P. W. 1993. Adaptation and the goals of evolutionary research. *Quarterly Review of Biology*, 68, 1–32.

Rennie, J. 2002. Fifteen answers to creationist nonsense. *Scientific American*, 287, 78–85.

Rose, K. D. 2001. The ancestry of whales. *Science*, 293, 2216–2217.

Schiestl, F. P., Ayasse, M., Paulus, H. F., Löfstedt, C., Hansson, B. S., Ibarra, F. & Francke, W. 1999. Orchid pollination by sexual swindle. *Nature*, 399, 421–422.

Schiestl, F. P., Ayasse, M., Paulus, H. F., Löfstedt, C., Hansson, B. S., Ibarra, F. & Francke, W. 2000. Sex pheromone mimicry in the early spider orchid (*Ophrys sphegodes*): Patterns of hydrocarbons as the key mechanism for pollination by sexual deception. *Journal of Comparative Physiology A*, 186, 567–574.

Schmid, R. & Schmid, M. J. 1977. Fossil history of the Orchidaceae. In *Orchid Biology: Reviews and Perspectives, I* (Ed. by Arditti, J.), pp. 25–45. Ithaca, NY: Cornell University Press.

Smithson, A. & Gigord, L. B. D. 2001. Are there fitness advantages in being a rewardless orchid? Reward supplementation experiments with *Barlia robertiana*. *Proceedings of the Royal Society of London, B* 268, 1435–1441.

Spomer, G. G. 1999. Evidence of protocarnivorous capabilities in *Geranium viscosissimum* and *Potentilla arguta* and other sticky plants. *International Journal of Plant Sciences*, 160, 98–101.

Sun, G., Ji, Q., Dilcher, D. L., Zheng, S. L., Nixon, D. C., & Wang, X. F. 2002. Archaefructaceae, a new basal angiosperm family. *Science*, 296, 899-904.

Thewissen, J. G. M. & Bajpai, S. Whale origins as a poster child for macroevolution. *BioScience*, 51, 1037–1049.

Thewissen, J. G. M., Williams, E. M., Roe, L. J. & Hussain, T. 2001. Skeletons of terrestrial cetaceans and the relationship of whales to artiodactyls. *Nature*, 413, 277–281.

Wasserthal, L. T. 1997. The pollinators of the Malagasy star orchids *Angraecum sesquipedale, A. sororium* and *A. compactum* and the evolution of extremely long spurs by pollinator shift. *Botanica Acta*, 110, 343–353.

Williams, N. H. 1982. The biology of orchids and euglossine bees. In *Orchid Biology: Reviews and Perspectives, II* (Ed. by Arditti, J.), pp. 119–171. Ithaca, NY: Cornell University Press.

Williams, S. E., Albert, V. A. & Chase, M. W. 1994. Relationships of Droseraceae: a cladistic analysis of *rbcL* sequence and morphological data. *American Journal of Botany*, 81, 1027–1037.

Wright, R. 2001. The "new" creationism. *Slate* (April 16), slate.msn.com.

Chapter 5: Orchids, Species, and Names

Blaxell, D. F. 1972. *Arthrochilus* F. Muell. and related genera (Orchidaceae) in Australasia. *Contributions from the N.S.W. National Herbarium*, 4, 275-283.

Bower, C. C. 1996. Demonstration of pollinator-mediated reproductive isolation in sexually deceptive species of *Chiloglottis* (Orchidaceae: Caladeniinae). *Australian Journal of Botany*, 44, 15–33.

Burke, J. M. & Adams, P. B. 2002. Variation in the *Dendrobium speciosum* (Orchidaceae) complex: a numerical approach to the species problem. *Australian Systematic Botany*, 15, 63–80.

Frost, J. S. & Platz, J. E. 1983. Comparative assessment of modes of reproductive isolation among 4 species of leopard frogs (*Rana pipiens* complex). *Evolution*, 37, 66–78.

Hopper, S. D. 2000. How well do phylogenetic studies inform the conservation of Australian plants? *Australian Journal of Botany*, 48, 321–328.

Hopper, S. D. & Brown, A. P. 2001. *Spider, Fairy and Dragon Orchids of Western Australia*. Bentley Delivery Centre, Western Australia: Department of Conservation and Land Management.

Hopper, S. D. & Brown, A. P. 2004. Robert Brown's *Caladenia* revisited, including a revision of its sister genera *Cyanicula*, *Ericksonella*, and *Pheladenia* (Caladeniinae: Orchidaceae). *Australian Systematic Botany*, 17, 171–240.

Jones, D. L., Clements, M. A., Sharma, I. K. & Mackenzie, A. M. 2001. A new classification of *Caladenia* R. Br. (Orchidaceae). *Orchadian*, 13, 389–419.

Koerner, L. 1999. *Linnaeus: Nature and Nation*. Cambridge, MA: Harvard University Press.

Kores, P. J., Molvray, M., Weston, P. H., Hopper, S. D., Brown, A. P., Cameron, K. M. & Chase, M. W. 2001. A phylogenetic analysis of Diurideae (Orchidaceae) based on plastid DNA sequence data. *American Journal of Botany*, 88, 1903–1914.

Mant, J. G., Schiestl, F. P., Peakall, R. & Weston, P. H. 2002. A phylogenetic study of pollinator conservatism among sexually deceptive orchids. *Evolution*, 56, 888–898.

Paetkau, D., Shields, G. F. & Strobeck, C. 1998. Gene flow between insular, coastal and interior populations of brown bears in Alaska. *Molecular Ecology*, 7, 1283–1292.

Roca, A. L., Georgiadis, N., Pecon-Slattery, J. & O'Brien, S. J. 2001. Genetic evidence for two species of elephants in Africa. *Science*, 293, 1473–1477.

Szlachetko, D. L. 2001. Genera et species Orchidalium 1. *Polish Botanical Journal*, 46, 11-26.

Sharma, I. K., Jones, D. L., Young, A. G. & French, C. J. 2001. Genetic diversity and phylogenetic relatedness among six endemic *Pterostylis* species (Orchidaceae; series Grandiflorae) of Western Australia, as revealed by allozyme polymorphisms. *Biochemical Systematics and Ecology*, 29, 697–710.

Wallace, L. E. 2003. The cost of inbreeding in *Platanthera leucophaea* (Orchidaceae). *American Journal of Botany*, 90, 235–242.

Chapter 6: Orchids, Biodiversity, and Hotspots

Beard, J. S., Chapman, A. R. & Gioia, P. 2000. Species richness and endemism in the Western Australian flora. *Journal of Biogeography*, 27, 1257–1268.

Brown, A., Thomson-Dans, C. & Marchant, N. (eds.) 1998. *Western Australia's Threatened Flora*. Como, Australia: Department of Conservation and Land Management.

Carroll, S. S. & Pearson, D. L. 1998. Spatial modeling of butterfly species richness using tiger beetles (Cicindelidae) as a bioindicator taxon. *Ecological Applications*, 8, 531–543.

Cowling, R. M. & Lamont, B. B. 1998. On the nature of Gondwanan species flocks: diversity of Proteaceae in Mediterranean south-western Australia and South Africa. *Australian Journal of Botany*, 46, 335–355.

Crisp, M. D., Laffan, S., Linder, H. P. & Monro, A. 2001. Endemism in the Australian flora. *Journal of Biogeography*, 28, 183–198.

Day, P. 2003. Watsonia control—a proven success. http://members.iinet. net.au/~ewan/watsonia__Peter.htm.

Hopper, S. D. 1979. Biogeographical aspects of speciation in the southwest Australian flora. *Annual Review of Ecology and Systematics*, 10, 399–422.

Lamont, B. B. & Connell, S. W. 1996. Biogeography of *Banksia* in southwestern Australia. *Journal of Biogeography*, 23, 295–309.

Myers, N., Mittermeier, R. A., Mittermeier, C. G., da Fonseca, G. A. B. & Kent, J. 2000. Biodiversity hotspots for conservation priorities. *Nature*, 403, 853–858.

Richardson, J. E., Weitz, F. M., Fay, M. F., Cronk, Q. C. B., Linder, H. P., Reeves, G. & Chase, M. W. 2001. Rapid and recent origin of species richness in the Cape flora of South Africa. *Nature*, 412, 181–183.

Thomson, C., Hall, G. & Friend, G. 1993. *Mountains of Mystery*. Como, Australia: Department of Conservation and Land Management.

Chapter 7: Orchids and Conservation

Batty, A. L., Dixon, K. W., Brundrett, M. & Sivasithamparam, K. 2001. Constraints to symbiotic germination of terrestrial orchid seed in a Mediterranean bushland. *New Phytologist*, 152, 511–520.

Clements, M. A. 1988. Orchid mycorrhizal associations. *Lindleyana*, 3, 73–86.

Curtis, A. & De Lacy, T. 1996. Landcare in Australia: does it make a difference? *Journal of Environmental Management*, 46, 119–137.

Finlayson, H. H. 1936. *The Red Centre*. Sydney: Angus & Robertson.

Flannery, T. F. 1995. *The Future Eaters*. New York: Braziller.

Hicks, A. J. 2000. *Asymbiotic Technique of Orchid Seed Germination*. Chandler, AZ: Orchid Seedbank Project.

Hobbs, R. J. 2001. Synergisms among habitat fragmentation, livestock grazing, and biotic invasions in southwestern Australia. *Conservation Biology*, 15, 1522–1528.

Lentic, M. 2000. Dispossession, degradation and extinction: Environmental history in arid Australia. *Biodiversity and Conservation*, 9, 295–308.

Lonsdale, W. M. 1994. Inviting trouble: Introduced pasture species in northern Australia. *Australian Journal of Ecology*, 19, 345–354.

Phillimore, R., Brown, A., Kershaw, K., Holland, E. & English, V. 2000. Elegant spider orchid (*Caladenia elegans* MS) interim recovery plan 2000–2003, pp. 1–15. Wanneroo, Western Australia: Department of Conservation and Land Management.

Rasmussen, H. N. 1995. *Terrestrial Orchids: From Seed to Mycotrophic Plant*. Cambridge, UK: Cambridge University Press.

Rauzon, M. J. & Drigot, D. C. 2002. Red mangrove eradication and pickleweed control in a Hawaiian wetland, waterbird responses, and lessons learned. In *Turning the Tide: The Eradication of Invasive Species* (Ed. by Veitch, C. R.). Auckland, New Zealand: IUCN.

Reinikka, M. A. 1995. *A History of the Orchid*. Portland, OR: Timber Press.

Sanderson, E. W., Jaiteh, M., Levy, M. A., Redford, K. H. Wannebo, A. V. & Woolmer, G. 2002. The human footprint and the last of the wild. *BioScience*, 52, 891–904.

Short, J. & Smith, A. P. 1994. Mammal decline and recovery in Australia. *Journal of Mammalogy*, 75, 288–297.

Simberloff, D. 2001. Eradication of island invasives: Practical actions and results achieved. *Trends in Ecology and Evolution*, 16, 273–274.

Sinden, J., Jones, R., Hester, S., Odom, D., Kalisch, C., James, R., & Cacho, O. 2004. The economic impact of weeds in Australia. *CRC for Australian Weed Management, Technical Series*, 8, 1-55.

Stewart, S. L. & Zettler, L. W. 2002. Symbiotic germination of three semi-aquatic rein orchids (*Habenaria repens, H. quinquiseta, H. macroceratitis*) from Florida. *Aquatic Biology*, 72, 25–35.

Wills, R. T. 1993. The ecological impact of *Phytophthora cinnamomi* in the Stirling Range National Park, Western Australia. *Australian Journal of Ecology*, 18, 145–159.

Chapter 8: Happy Hunting

Bates, R. J. & Weber, J. Z. 2000. *Orchids of South Australia*. South Australia: Government Printer.

Backhouse, G. N. & Jeanes, J. A. 1995. *The Orchids of Victoria*. Carlton, Victoria: Melbourne University Press.

Bentley, S. 2000. *Native Orchids of the Southern Appalachian Mountains*. Chapel Hill: University of North Carolina Press.

Bishop, T. 2000. *Field Guide to the Orchids of New South Wales and Victoria*. Sydney, New South Wales: University of New South Wales Press.

Brown, P. M. 2003. *Wild Orchids of North America, North of Mexico.* Gainesville: University of Florida Press.

Coleman, R. A. 1995. *The Wild Orchids of California.* Ithaca, NY: Cornell University Press.

Coleman, R. A. 2002. *The Orchids of Arizona and New Mexico.* Ithaca, NY: Cornell University Press.

Correll, D. S. 1978. *Orchids of North America, North of Mexico.* Palo Alto, CA: Stanford University Press.

Jones, D., Wapstra, H., Tonelli, P., & Harris, S. 1999. *The Orchids of Tasmania.* Carlton, Victoria: Melbourne University Press.

Keenan, P. 1998. *Wild Orchids Across North America.* Portland, OR: Timber Press.

Luer, C. A. 1975. *The Native Orchids of the United States and Canada, excluding Florida.* New York: New York Botanical Garden.

Peterson, R. T. & McKenny, M. 1968. *A Field Guide to the Flowers: Northeastern and North-Central North America.* New York: Houghton Mifflin.

St. George, I., Irwin, B., Hatch, D., & Scanlen, E. 2001. *Field Guide to New Zealand Orchids.* Wellington: New Zealand Native Orchid Group.

Index

30. Spellot

Daedalus 5.58 101743